AF400830

43   # An Introduction to Immersed Boundary Methods

The *Cambridge Monographs on Applied and Computational Mathematics* series reflects the crucial role of mathematical and computational techniques in contemporary science. The series publishes expositions on all aspects of applicable and numerical mathematics, with an emphasis on new developments in this fast-moving area of research.

State-of-the-art methods and algorithms as well as modern mathematical descriptions of physical and mechanical ideas are presented in a manner suited to graduate research students and professionals alike. Sound pedagogical presentation is a prerequisite. It is intended that books in the series will serve to inform a new generation of researchers.

A complete list of books in the series can be found at
www.cambridge.org/mathematics.

Recent titles include the following:

27. Difference equations by differential equation methods, *Peter E. Hydon*
28. Multiscale methods for Fredholm integral equations, *Zhongying Chen, Charles A. Micchelli & Yuesheng Xu*
29. Partial differential equation methods for image inpainting, *Carola-Bibiane Schönlieb*
30. Volterra integral equations, *Hermann Brunner*
31. Symmetry, phase modulation and nonlinear waves, *Thomas J. Bridges*
32. Multivariate approximation, *V. Temlyakov*
33. Mathematical modelling of the human cardiovascular system, *Alfio Quarteroni, Luca Dede', Andrea Manzoni & Christian Vergara*
34. Numerical bifurcation analysis of maps, *Yuri A. Kuznetsov & Hil G.E. Meijer*
35. Operator-adapted wavelets, fast solvers, and numerical homogenization, *Houman Owhadi & Clint Scovel*
36. Spaces of measures and their applications to structured population models, *Christian Düll, Piotr Gwiazda, Anna Marciniak-Czochra & Jakub Skrzeczkowski*
37. Quasi-interpolation, *Martin Buhmann & Janin Jäger*
38. The Christoffel–Darboux kernel for data analysis, *Jean Bernard Lasserre, Edouard Pauwels & Mihai Putinar*
39. Geometry of the phase retrieval problem, *Alexander H. Barnett, Charles L. Epstein, Leslie Greengard & Jeremy Magland*
40. Discrete variational problems with interfaces, *Roberto Alicandro, Andrea Braides, Marco Cicalese & Margherita Solci*
41. Spectral and spectral element methods for fractional ordinary and partial differential equations, *Mohsen Zayernouri, Li-Lian Wang, Jie Shen & George Em Karniadakis*
42. High-accuracy finite difference methods, *Bengt Fornberg*

An Introduction to Immersed Boundary Methods

ROBERTO VERZICCO
Università degli Studi di Roma "Tor Vergata"
Gran Sasso Science Institute, L'Aquila
University of Twente, Enschede

MARCO D. DE TULLIO
Politecnico di Bari

FRANCESCO VIOLA
Gran Sasso Science Institute, L'Aquila

CAMBRIDGE
UNIVERSITY PRESS

Shaftesbury Road, Cambridge CB2 8EA, United Kingdom

One Liberty Plaza, 20th Floor, New York, NY 10006, USA

477 Williamstown Road, Port Melbourne, VIC 3207, Australia

314–321, 3rd Floor, Plot 3, Splendor Forum, Jasola District Centre,
New Delhi – 110025, India

103 Penang Road, #05–06/07, Visioncrest Commercial, Singapore 238467

Cambridge University Press is part of Cambridge University Press & Assessment,
a department of the University of Cambridge.

We share the University's mission to contribute to society through the pursuit of
education, learning and research at the highest international levels of excellence.

www.cambridge.org
Information on this title: www.cambridge.org/9781009123204
DOI: 10.1017/9781009128438

When citing this work, please include a reference to the DOI 10.1017/9781009128438

First published 2026

A catalogue record for this publication is available from the British Library

*A Cataloging-in-Publication data record for this book is available from the Library of
Congress*

ISBN 978-1-009-12320-4 Hardback

"Everyone knew it was impossible, until a fool who didn't know came along and did it." A. Einstein

Contents

Preface

Immersed boundary methods (IBMs), broadly speaking, are a class of numerical techniques that enable the solution of partial differential equations over domains with complex curvilinear boundaries using simple, usually orthogonal, non-body-conformal meshes.

Despite the seeming simplicity of IBM implementation in numerical codes, the necessary fine tuning of several algorithmic details entails considerable work and a comprehensive grasp of the method, which can hardly be achieved through the reading of research articles.

In fact, whereas there is a vast and continuously expanding literature on IBMs, no sources provide a systematic introduction to the subject. As a result, the newcomers of this growing community might easily get lost in the wealth of publications which, by necessity, focus either on particular aspects of a single technique or on specific applications.

This book aims to fill this gap by explaining all the components of the most common IBMs for computational fluid dynamics (CFD) in order to discuss the advantages and drawbacks of the different approaches.

The book starts with a brief motivation for the employment of IBMs and then provides a historical overview of the subject in CFD, from the hidden origins up to the beginning of the new millennium, when the number of papers has literally exploded.

Although the book is intended for students and researchers who will use it in conjunction with CFD software, an effort has been made to make it self-contained. Accordingly, in Chapter 2, the governing equations for fluid dynamics are derived, and some fundamental concepts of their numerical integration are introduced.

Chapter 3 opens the technical section related to IBMs, discussing the description of complex geometry objects and the tagging of the mesh nodes with respect to the object itself. Chapter 4 describes different approaches to alter the governing equations to prescribe the appropriate conditions on boundary nodes.

Chapters 5 to 7, respectively, extend the concepts to moving boundaries, explain how to obtain smooth distributions of the hydrodynamic loads and discuss the additional difficulties introduced by fluid–structure-interaction.

As the traditional CFD bottlenecks of near-wall resolution and turbulence modeling are exacerbated in the IBM context, several possible extensions are illustrated in Chapter 8 to enable the application of IBMs to high Reynolds number flows.

Chapter 9 concludes the theoretical part with some examples of IBMs in advanced physical problems; they are illustrated with the aim of providing additional suggestions for a wider application of the techniques introduced in the previous chapters.

Finally, Chapter 10 is devoted to the application of IBMs to numerical examples. We describe the structure of a computer code developed and extensively tested for this purpose, which can be freely downloaded from the web: `https://gitlab.com/vdv9265847/IBbookVdV`. In this flow solver, some of the IBMs discussed in the book are already implemented and they can be changed simply by switching flags in the input file. Several numerical applications of increasing complexity, ranging from two-dimensional flows around rigid fixed bodies up to complex three-dimensional flows with fluid–structure interaction are then presented; they can be duplicated autonomously by the interested reader and used as benchmarks.

We believe that with the present structure, the book serves a twofold purpose: (i) it may guide beginners step-by-step to design and realize the IBM implementation that better meets their needs, and (ii) it provides a comprehensive discussion of the expanding field of IBMs within a unified context.

The first input about IBM came from Professor Paolo Orlandi, the mentor of one of the authors, who in 1997 suggested the reading of a freshly published paper which might have contained interesting material. Even if it turned out that the IBM therein was limited in many aspects, that suggestion lit the spark triggering all the successive developments.

During the 1998 Summer Program of the Center for Turbulence Research at Stanford University, Professor Parviz Moin gave RV the chance to test his ideas during a one-month project despite the comment from another renowned participant who said "it can't work!". The final results have proven otherwise and the reported quotation by A. Einstein is along this line.

Acknowledgments

The project of this book was conceived a long time ago, when Professor Gianluca Iaccarino, at the time PhD student of one of the authors, completed his thesis. Although his new research interests and the bright career have prevented him from being a coauthor, his contribution is gratefully acknowledged.

The authors wish to thank Professor Simone Camarri for a thoughtful reading of the manuscript. The precious help of the PhD students Giovanni Vagnoli, Bálint Nagy and Gianlorenzo Granata, and of the postdoc Valerio Lupi, whose comments have helped us to improve the clarity of the book, is also gratefully acknowledged.

Special thanks go to David Tranah of Cambridge University Press, who has been wonderfully encouraging and extremely patient with us.

1

Introduction

1.1 Motivation

Computational science is a relatively new, interdisciplinary approach to the study of natural and technological complex systems that combines elements from different scientific fields and relies on computer-based models.

The quote "The computer is incredibly fast, accurate, and stupid. Man is incredibly slow, inaccurate, and brilliant. The marriage of the two is a force beyond calculation" has been attributed, with several variations, to many different scientists although Garland (1982) reports a specific event in which Leo Cherne wrote it in 1977: Regardless the original author, on account of the five-decade unrestrained growing power of computers, it sounds like a prophecy that has now come true.

In fact, alongside theory and laboratory experiments, numerical simulations have emerged as an essential tool for advancing science and technology with computer models that are routinely used for the analysis of complex systems.

Numerical computations are essential for the design and optimization of not only everyday products like cars, aircraft, buildings and bridges, but also innovative biomedical and microelectronic devices. Computational science is at the core of meteorology, oceanography, climatology and even of epidemiology (Smolinski et al., 2003) or dynamics of crowds and populations (Curtis et al., 2016).

The common feature of all these disciplines is the use of complex computational models to predict the system dynamics. These models can be fully exploited only by implementing dedicated numerical methods on powerful computers.

The main benefits of computation over other approaches are that system dynamics can be predicted in advance, hazardous scenarios can be safely explored, performances can be assessed and improved without employing hardware models, and interactions among various systems can be reproduced and studied in a controlled environment.

One of the branches of computational science that has greatly benefited from the availability of powerful computers is computational fluid dynamics (CFD). Traditionally, it has been the most demanding in terms of computational power.

The great scientist J. von Neumann was prophetic in this respect when in 1940 he wrote: "Our present analytical methods seem unsuitable for the solution of important problems arising in connection with non-linear partial differential equations ... The truth of this statement is particularly striking in the field of fluid dynamics" (Goldstine, 1980). In fact, the governing equations of fluid dynamics are quadratically nonlinear and strongly coupled, making analytical solutions unattainable in realistic conditions. This explains why using simplified models or empirical correlations limits the overall reliability of the results. By contrast, in sectors such as aeronautical or automotive engineering, where efforts have been made to develop and apply high-fidelity CFD models, the progress has been very rapid and this has prompted other branches of science to pursue a similar approach.

A growing number of engineers and scientists nowadays rely on high-fidelity CFD tools for the design and analysis of new technological solutions or complex systems. However, in spite of its widespread use, there are still difficulties that prevent CFD from being used as a standard tool. Among many, accurate predictions of highly turbulent flows, parametric analyses of complex systems and design optimization are examples of critical applications.

As an example, we can consider the flow around an object, such as the hull in Figure 1.1, having a complex geometry. As we will detail in Chapter 2, owing to the nonlinear nature of the governing relations, the only way to obtain a solution is to integrate the relations by a numerical technique, which requires a discretization procedure aimed at transforming the original set of differential equations into a discretized algebraic system. This implies that the entire domain of interest is tessellated by small elements, the so-called computational mesh, whose "distribution" and "quality" will determine the reliability of the numerical solution.

Usually starting from some geometrical description of the boundaries (bodies and external domain), a surface grid is first constructed to be used as a boundary condition for the generation of a volume mesh covering the whole fluid domain. Building the initial surface grid can be extremely difficult and time-consuming for realistic, complex boundaries, but the problem is further exacerbated when there are multiple moving elements (see Figure 1.1).

Although different techniques (structured or unstructured grids, block or overlapping meshes, see Thompson et al. (1999) for more details) are available for mesh generation, their common objective is to produce a *regular* discretization upon which the quality of the flow solution heavily relies. Regularity in this context is intended in general terms, for example, cell size and shape, element orthogonality, smooth variation of the spacing, etc.

In general, generating a high-quality grid, fitted to a complex three-dimensional object (as in Figure 1.1), can become difficult enough that only highly trained and specialized researchers accomplish it. Furthermore, the procedure can be so time-consuming that it can easily exceed the time needed to obtain the flow solution. Usually, the resulting meshes are typically not guaranteed to be orthogonal, and this requires more complex solution algorithms with significant overhead in the per-cell operation count and, often, with high computational cost and limited precision (Chung, 2002).

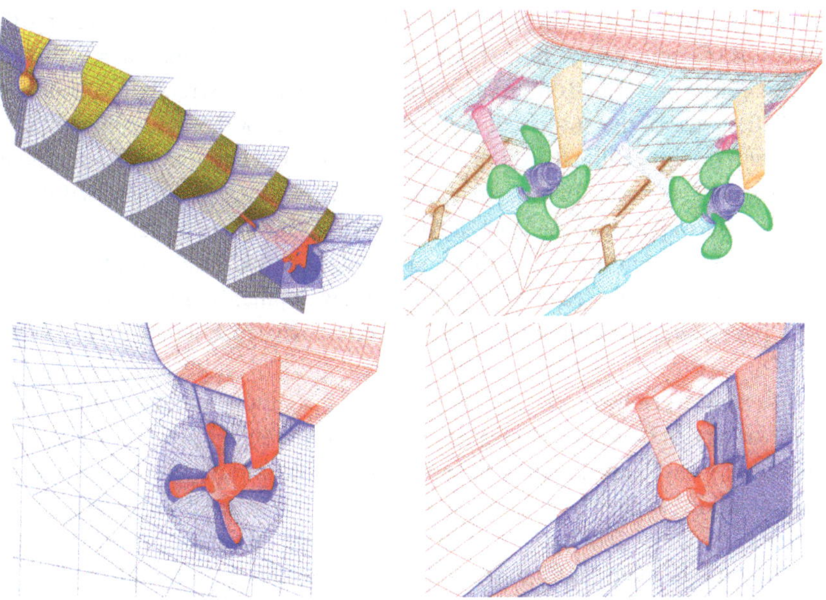

Figure 1.1 Example of body-fitted overset grids around a hull with appendices and impellers. Courtesy of Professor A. Di Mascio, reproduced with permission.

In some applications, the boundaries are not even stationary and when there are multiple parts whose relative position varies, simply changing the reference frame is not enough to compensate for the motion. This entails additional complexity since solution methods that rely on body-conformal discretizations require complex algorithms such as sliding surfaces, overlapping grids and deformation or regeneration of the mesh. Obviously, all these procedures need interpolations of the solution, thus resulting in a degradation of accuracy and conservation properties, in addition to the considerable overhead of computational time.

In this context, it is clear that a numerical procedure capable of handling complex geometric configurations without resorting to body-fitted meshes would contribute significantly to promoting the application of CFD to complex problems: The **immersed boundary method** (IBM) has emerged in the past two decades as a suitable and versatile candidate.

1.2 Background

The key idea of the immersed boundary (IB) approach is to replace the boundaries by a time–space varying distribution of force densities mimicking the effect of the body on the flow. The main advantage of this approach is that the forces can be prescribed on simple and regular meshes, thus retaining all the ease and efficiency of the numerical methods developed for these discretizations.

It will be shown that there is a wide variety of possible forcing terms and an even wider assortment of implementations, thus yielding a multitude of different IBMs, each one with advantages and drawbacks. The aim of this book is to guide the reader through the basics of the methodology and to the implementation of some of the main techniques; the latest developments are left to the scientific papers whose number has grown continuously since the beginning of the 2000s.

The immersed boundary technique allows the simulation, using simple discretizations, of flows in complex geometries by forcing the governing equations along surfaces corresponding to the physical location of the boundaries; the computation can then be performed on a much simpler mesh with fewer computational resources. The essence of the method is the decoupling of the geometry of the fluid domain and the grid for the flow solution. In more detail, if the flow around object $\mathbf{B}$ in Figure 1.2 has to be computed in the fluid domain Γ, a standard CFD approach would require the generation of a grid in the region bounded by $\partial\mathbf{B}$ and $\partial\Gamma$; a possible standard configuration is that of Figure 1.2a. In IBM, in contrast, an underlying grid is laid in the inner region of $\partial\Gamma$ regardless of the shape of the object $\mathbf{B}$, which is *immersed* into the computational domain. The effect of the body on the flow is enforced by prescribing a suitable distribution of fictitious terms (forces, velocities or generalized constraints) such that the correct boundary conditions on $\partial\mathbf{B}$ are obtained for the flow.

Of course, in the IB context $\partial\mathbf{B}$ is not a coordinate line (see Figure 1.3), so the computational cells must be classified according to their position relative to the

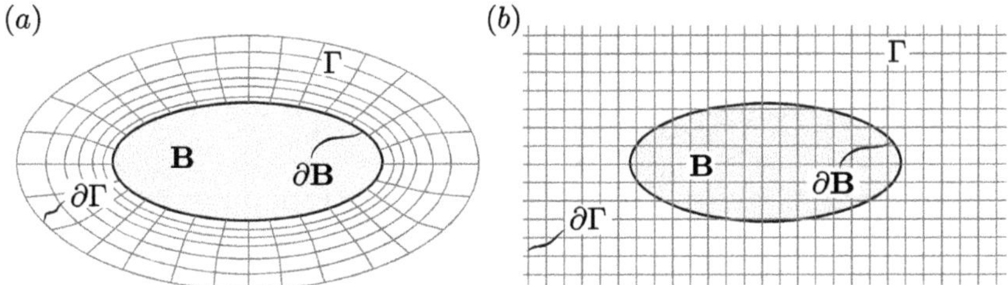

Figure 1.2 Sketch of the domain discretization with a body-conformal mesh (a) and with the immersed boundary method (b).

body ("internal," "external" and "interface"); and while ad hoc strategies can be devised for simple objects, for general complex geometries the task can become computationally intensive. Another difficulty is that the spatial position of the discrete unknowns generally does not coincide with the body surface where the former are known through the boundary conditions. This introduces the additional challenge of forcing the equations at some nodes to impose a solution at a different position, which involves appropriate interpolation techniques.

These problems become more severe when the object is moving or, even worse, deforming with respect to the grid, since the steps of cell tagging and interpolation/reconstruction at the IB must be repeated at every time step. In addition, when a body moves across the grid, there are cell nodes whose state changes in one time step from "internal" to "interface" and they miss all the necessary flow history to compute its evolution.

Each of the above difficulties has been tackled by many different research groups worldwide and many ingenious numerical algorithms have been made available through scientific publications. If we recall that, in fact, the body is not physically present in the domain except for some layers of forced computational nodes, we could simplify the essence of the technique by stating that immersed boundary methods consist of the "art" of imposing boundary conditions in the middle of the computational domain; though at first glance the principle might seem hopeless, it is in fact a very effective strategy and showing the advantages and weaknesses of these methods is the main aim of this book.

1.3 Literature and History

The idea of using non-body-conformal meshes in computational fluid dynamics is definitely not new and it has been pursued by many researchers at least since the 1950s. The *father* of immersed boundaries is unanimously considered to

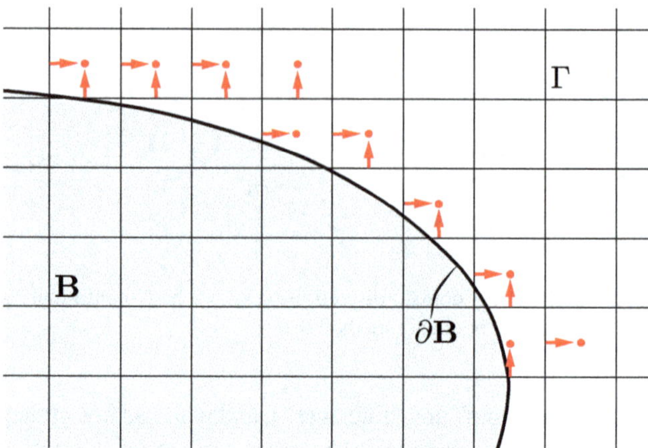

Figure 1.3 Detail of the intersection between an immersed boundary and the grid; the red arrows and bullets indicate possible positions of the discretized unknowns (on a staggered mesh) with respect to the boundary.

be Charles Peskin who, in his seminal paper (Peskin, 1972a), showed the huge potential of the method by simulating the two-dimensional flow through the natural mitral heart valve. The basic concepts, however, had already been around for a while and, to the authors' knowledge, the first published IB example was given by Gentry et al. (1966) for unsteady compressible flows. In this paper, the authors note that the governing "equations for those cells adjacent to the bounding surface of a rigid body are modified to insure that there is no flow of mass or energy across the boundary." The backbone of the proposed numerical scheme, however, was the "fluid-in-cell" method (Rich, 1963) in turn relying on the work of Evans and Harlow (1957): Already in these reports some techniques for computing compressible flows around obstacles without using body-fitted meshes were described in detail and numerical calculations on two-dimensional Cartesian meshes with up to a few thousand nodes were shown. We report here, as a historical note, some sketches taken from those reports (Figure 1.4) where it can be appreciated how the same concepts as those in Figure 1.3 had already been put forward in the middle of the 1950s. As evident from Figure 1.4c, not only boundaries aligned with the Cartesian mesh were considered, but also surfaces that could intersect the coordinate lines with arbitrary position and orientation.

Incompressible flows were tackled using a similar approach (see Welch et al., 1965) and in the paper by Harlow and Whelch (1971) the authors presented the marker and cell (MAC) method in which two-dimensional flows

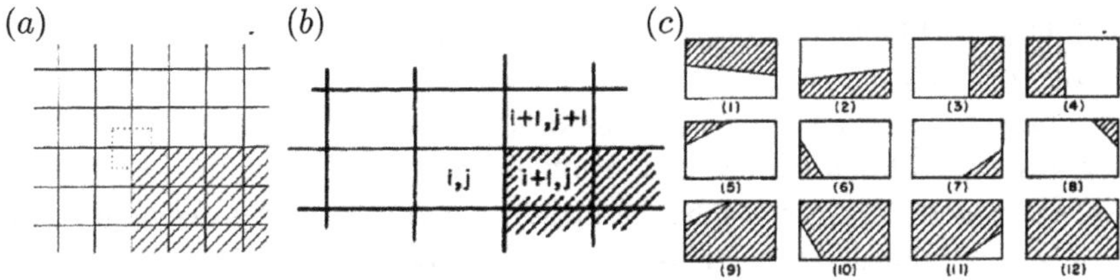

Figure 1.4 Schematics of boundary–mesh intersection extracted from: (a) Evans and Harlow (1957), p. 26; (b) Rich (1963), p. 36; (c) Rich (1963), p. 38.

with free surface and obstacles were computed. In the simulations, the Eulerian (Cartesian) grid was not fitted to the obstacles and the nodes of the mesh were tagged as fluid or solid to impose the appropriate velocity boundary conditions through finite differences. Although the main idea of an immersed boundary was already there, the authors stated that "walls are restricted in orientation so that they lie along the boundaries of the Eulerian cells. Relaxation of this requirement could be accomplished only at the expense of a considerable increase in complication."

The above limitation was removed a few years later by Viecelli (1969), who extended the MAC method to include boundaries of arbitrary shape. The basic idea consisted of treating the fluid–boundary interface as a free surface and imposing there pressure boundary conditions so that particles could move only tangentially to the surface. This procedure led to an iteration between pressure and velocity fields until flow incompressibility and boundary impermeability were both satisfied. The method, referred to as ABMAC (arbitrary boundary MAC), was also generalized in a follow-up paper (Viecelli, 1971) for handling moving walls and, in this case, velocity boundary conditions, in addition to pressure, were imposed at the interface. This technique allowed the treatment of walls moving with a prescribed law or as a consequence of the forces exerted by the fluid on the surface. Early attempts to solve partial differential equations on non-body-conformal meshes can also be found in the scientific literature of countries in the former Soviet bloc. The diffusion equation $\nabla \cdot (k\nabla\Phi) = f$, with k a position-dependent diffusivity of the scalar field Φ, was solved by Saulev (1963) on a nonfitted mesh by introducing a modified diffusivity $k_c = k(1 + 1/\epsilon)$ with $\epsilon \to \infty$ in the fluid domain and $\epsilon \to 0$ inside the body. Rigorous proof was provided that the resulting solution Φ_c approached the correct one Φ in the limiting behavior. Kuznetsov (2001) later showed that approach to be equivalent to introducing a Lagrangian multiplier $\lambda = \Phi/\epsilon$ in the region inside the body and letting $\epsilon \to 0$: This is essentially the distributed

Lagrange multiplier fictitious domain method proposed by Dihn et al. (1992) thirty years later.

Yanenko (1967) formulated a method, similar to that of Gentry et al. (1966), for solving fluid dynamics equations on non-body-conformal meshes referred to as the "unmatched or incoherent grids" (Figure 1.5b); after explaining how to use a fractional-step method to solve flow problems on body-fitted grids as in Figure 1.5a, he explicitly states that "for incoherent grids additional interpolations are needed in order to impose boundary conditions at the walls."

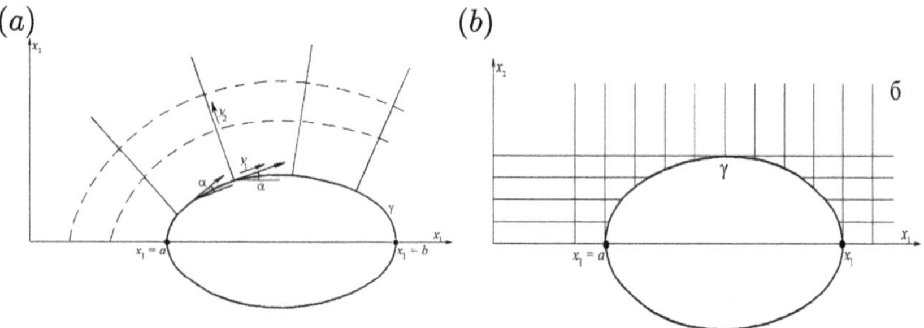

Figure 1.5 Representation of "matched" (a) and "unmatched" (b) grids for the flow around a curvilinear object as represented by Yanenko (1967) on p. 130 and p. 137 of his book.

At the beginning of the 1970s, Peskin (1972a) reported simulations of the blood flow in the heart system assuming low Reynolds number and two-dimensional dynamics. The approach was innovative since the incompressible Navier–Stokes equations were solved on a Cartesian grid but the boundaries were discretized by Lagrangian markers connected by elastic elements *immersed* in the flow: Fluid and boundaries exerted forces on each other thus determining the flow and the structure dynamics. The Lagrangian markers moved with the local fluid velocity and their location was tracked in space; the position of the markers and their forcing on the fluid were then transferred to the Eulerian mesh, where the flow solution was obtained. The method was extended in Peskin (1972b) to mimic the active contraction of the heart muscular fibers, relying on a clever set of links among the Lagrangian markers; using a mesh discretized by 64^2 nodes (and low Reynolds number), the basic left-heart flow features were reproduced numerically.

Benefiting also from a continuous growth of available computational power, the procedure was further improved by Peskin and McQueen (1989) and by McQueen and Peskin (1989), who were able to simulate the three-dimensional blood flow in the heart on a 128^3 cubic lattice using realistic contractile fibers.

Although the method was originally developed to simulate flows interacting with elastic boundaries, in principle it could also be used for rigid objects once the links connecting the Lagrangian markers were made stiff enough to prevent the structure from deforming. Unfortunately, this strategy is inefficient as it leads to stiffer equations that require significant computational resources for integration; we will come back to this point in the later chapters when various IB implementations will be illustrated in detail.

A different IB approach for flows around solid, rigid objects was proposed by Basdevant and Sadourny (1983) for the incompressible Navier–Stokes equations in vorticity–streamfunction formulation and later extended by Briscolini and Santangelo (1989) to primitive variables (velocity and pressure). These authors note that the procedure "treats the no-slip condition of the Navier–Stokes equations as a forcing term acting, at each time step, on the physical boundary ∂B; the forcing depends on the velocity field." The numerical scheme was referred to as the "mask method" since it consisted essentially of deleting the field inside the body by a mask, and smoothly connecting it to the external field.

They computed the unsteady two-dimensional flow around circular and square cylinders at Reynolds numbers up to 1000, obtaining a good agreement with the results from the literature.

Rather than altering the velocity field inside the body, Goldstein et al. (1993) exploited the body force term of the Navier–Stokes equations; "forces were chosen to lie along a desired surface and to have a magnitude and a direction opposing the local flow such that the flow was brought to rest on an element of the surface." This resulted in an easy implementation of the method that was flexible enough to allow the simulation of the three-dimensional plane- and ribbed-turbulent channel flow.

The main drawback of the method of Goldstein et al. (1993) is that it contains free parameters requiring a problem-dependent tuning; in particular, for unsteady flows this forcing introduces a time step limitation which reduces the efficiency and the applicability of the technique (see the following chapters for a detailed description).

Soon after, Saiki and Biringen (1996) found that the body forces proposed in Goldstein et al. (1993) were more efficient when used in combination with finite-difference discretizations than with spectral methods. The reason was that the immersed boundaries forcing terms, localized at the fluid–boundary interface, tend to generate unphysical oscillations when described by nonlocal spectral operators (Gibbs phenomenon).

Consistent results were obtained by Elghaoui and Pasquetti (1996) that applied a spectral embedding method to a convection diffusion equation. They used Fourier expansions for the latter and a boundary integral equation to enforce the boundary conditions on a hexagonal line in a two-dimensional plane. However, because of the discontinuities introduced by the boundaries, the convergence rate of the method was not fully satisfactory and the authors concluded that the accuracy of the solution in the vicinity of the boundaries needed to be improved.

Cortez (2000) and Cortez and Minion (2000) proposed a variant of the Peskin approach in which the momentum equation was forced through a series of regularized vortex dipoles (referred to as vortex blobs) that were shown to be more accurate than standard forcings. The numerical examples, however, were limited to two-dimensional flows at relatively low Reynolds numbers.

Despite the increasing number of papers dealing with or applying immersed boundary techniques, until the late 1990s, the scenario was essentially dominated by the method developed by Peskin (1972a); the review by Peskin (2002) is an exhaustive reference for the state of the art at the time.

An important breakthrough was obtained by Mohd-Yusof (1997) who derived an alternative formulation of the forcing based on the time-discrete version of the Navier–Stokes equations that neither requires user-defined parameters nor affects the stability of the time integration; these are highly desirable features in order to make the approach flow independent and the simulation of complex three-dimensional flows feasible. Mohd-Yusof combined the new method with B-splines to compute the laminar flow over a three-dimensional ribbed channel, showing substantial improvements with respect to the earlier formulations.

A forcing based on the idea of Mohd-Yusof (1997) was developed by Fadlun et al. (2000) using finite differences and extended in Verzicco et al. (2000) to large-eddy simulation; in both cases, second-order accuracy was preserved in the IB context. The applications included the flow around simple and complicated geometries for a wide range of Reynolds numbers.

The review papers by Iaccarino and Verzicco (2003) and Mittal and Iaccarino (2005) provide additional references and a comprehensive discussion of almost all methods developed until that period, and the advantages and drawbacks of the various techniques are illustrated through specific numerical examples and applications. Since then, however, the number of IB papers has literally exploded and accounting for all possible developments and applications has become practically impossible. In Figure 1.6, we report the number of papers per year published in the period 1970–2020 and containing the words "immersed boundary" in the title; it can be seen that while initially these publications were only sporadic, since 2000 they have constantly grown at a rate that has been sustained. It is also worthwhile noting that Figure 1.6 does not account for all the literature on the subject since expressions like "virtual boundary," "immersed surface," "boundary body forces," "mask method" and similar have not been included in the search.

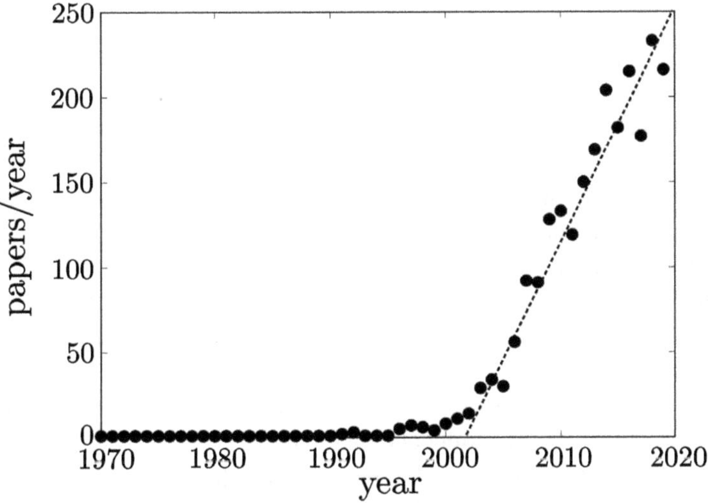

Figure 1.6 Number of published papers per year in the period 1970–2020 and containing in the title the words "immersed boundary" or "immersed boundaries" in all fields. The fit in the range 2000–2020 indicates a growth with a constant acceleration of 10 papers/year^2 since the year 2000. The dashed line is the fit papers=5*(year-2000)^2. Source: Scopus, December 2022.

Even if we are conscious that a comprehensive literature review covering the past two decades is not possible, we wish to mention those contributions that have fostered innovative research in IBMs.

The Physalis method (Prosperetti and Oğuz, 2001; Zhang and Prosperetti, 2005) is somehow unique in the field of IB since it can simulate flows with spherical particles using Cartesian grids, although it cannot be easily generalized to different complex geometries. In fact, the main idea is that, owing to the no-slip condition, in the reference frame of each sphere the near-wall velocity is so small that the Stokes equations are an excellent approximation to the full Navier–Stokes system, and this is true regardless of the large-scale Reynolds number of the external flow. Since the Stokes flow around a sphere is known analytically, any information can be transferred to the surrounding Eulerian points without approximations, and the forces can also be computed accurately. However, the analytical solution for the Stokes problem is not known for a generic complex geometry and the method loses all advantages.

In the context of incompressible flows, Kim et al. (2001) observed that using IB forces in combination with a fractional-step integration scheme (Kim and Moin, 1985) could yield a velocity field that was locally non-solenoidal. They solved the problem by also forcing the continuity equation by source/sink terms at the cells crossed by the immersed boundary, thus eliminating the mass conservation problem. The proposed strategy was easy to implement and was quickly adopted by many research groups.

Tseng and Ferziger (2003) introduced ghost cells at the immersed interface to add flexibility to the imposition of boundary conditions; it was shown that while retaining the second-order accuracy, it was possible to impose Dirichlet, Neumann and Robin velocity boundary conditions.

All the previously described implementations of IBMs were based on flow solvers that relied on finite differences, finite volumes or spectral spatial discretizations. In Feng and Michaelides (2004) a lattice–Boltzmann method was instead coupled with the IB forcing of Peskin (1972a) in order to combine the most advantageous features of the two techniques. The resulting method could accurately reproduce the sedimentation of two circular particles in a quiescent fluid and the same phenomenon for more than 500 particles: All the cases were simulated in two dimensions and for low values of the Reynolds number.

Although the IB community had always considered the progeny of Peskin's methods (Peskin, 2002) and those derived by Mohd-Yusof (1997) as alternatives, Uhlmann (2005) showed that it was possible to combine the two to obtain a powerful scheme that allowed a smooth transfer between Eulerian and Lagrangian meshes (thus giving regular load distributions) while at the same time avoiding strong restrictions on the time step. The resulting method was efficient and accurate enough to allow the simulation of $O(1000)$ fully resolved spherical particles settling in a fluid at a particle Reynolds number of a few hundred, which was a landmark achievement in the field.

One of the applications in which IBMs really show all their potential is when multiple bodies are in relative motion, since this can be accounted for without deforming or regenerating the grid. Gilmanov and Sotiropoulos (2005) and Yang and Balaras (2006), however, noted that when a body moves, some computational nodes that at one time step are inside the body can become fluid points at the next step and they do not have a valid "fluid history" to compute all the terms of the Navier–Stokes equations. The problem was solved by a field extension procedure in which all the missing information was obtained by extrapolation from the closest fluid points.

Using finite-element approximations, Glowinski et al. (1998) resorted to distributed Lagrange multipliers to impose boundary conditions, within the fictitious domain method, to rigid bodies moving along a prescribed trajectory in an incompressible viscous fluid. Taira and Colonius (2007) observed that pressure also acts like a Lagrange multiplier to ensure incompressibility, and so the distributed IB forcing terms can be handled in a unified way in a projection method like the fractional step of Kim and Moin (1985). The method was shown to predict correctly some two-dimensional canonical flows without resorting to the additional forcing of the continuity equation of Kim et al. (2001) or to iterations between momentum forcing and pressure correction as suggested by Fadlun et al. (2000).

A versatile and powerful tool was developed by Ge and Sotiropoulos (2007) that combined a curvilinear-mesh body-fitted flow solver with a sharp interface IBM. This idea is very interesting since the curvilinear grid can be used to resolve the near-wall regions of a complex domain, while IBs can deal with small geometrical details or moving/deforming objects. The method was, in fact, used to simulate the flow through an aortic root with a mechanical heart valve in which the leaflet motion was determined by the hydrodynamic loads. The problem was studied further in a subsequent paper of Borazjani et al. (2008) that provided a theoretical framework for the application of IBMs to fluid–structure interaction problems.

A lot of effort in developing IBMs is made in reconstructing the solution at the immersed interface and in transferring information from the Eulerian mesh to the boundary element (or its Lagrangian counterpart) and vice versa. Starting from the ideas of Uhlmann (2005), Vanella and Balaras (2009) developed a moving least squares procedure to build the transfer function between the grid and the boundary. Among the advantages, there is the possibility to impose a variety of different boundary conditions and to compute analytically the solution derivatives at the wall through the basis functions. This method has been widely adopted in IB applications owing to its second-order accuracy and the smooth distribution of surface forces, which is mandatory when simulating deformable boundaries.

Although the vast majority of IBMs have been developed for and applied to incompressible flows, there have been attempts to employ the technique to compressible flows also. For example, de Tullio et al. (2007) combined IBMs with a local grid refinement procedure to simulate supersonic flows around spheres and transonic flows around airfoils, showing the full second-order accuracy.

Before concluding this literature review, we wish to point out that the techniques we have described can all be gathered under the generic name of immersed boundary methods, which are the main object of the present book. There are other procedures, however, that rely on a similar philosophy but differ in technical aspects. For the sake of brevity, we will not discuss such approaches in detail, although we now mention the main references in order to provide the interested reader with additional material.

A class of methods, referred to as "penalty methods" or "fictitious domain methods" or "domain embedding methods," considers the immersed body as a porous medium and solves the Navier–Stokes–Brinkman equations; these are the Navier–Stokes equations with the addition of a volumetric loss term, called Darcy drag, which accounts for the action of the porous medium on the flow. When the permeability of the material tends to zero, it behaves as a solid and the flow around an arbitrary shaped object can be simulated by prescribing this velocity penalization term in space. A detailed description of the method can be found in Heinrich and Vionnet (1995), Angot et al. (1999) and Khandra et al. (2000) and references therein. Penalty methods could also be considered, from a mathematical viewpoint, as a particular case of the IB forcing proposed by Goldstein et al. (1993) and Saiki and Biringen (1996), although the physical interpretation of the results is different. This point will be further discussed in the later chapters when we deal with the details of IB implementations.

A different technique, called the "Cartesian grid method," also relies on non-body-fitted meshes to describe the flow around complex geometries. In this case, however, instead of adding forcing terms to the governing equations, the grid cells are modified at the body interface according to their intersections with the underlying grid. Given the large number of possible intersections, this generates a wide variety of "cut interface cells," which must be handled in a separate way depending on their topology. Finite-volume methods have been successfully used for the calculation of complex three-dimensional flows, although most of the developments concerned the Euler equations (Epstein et al., 1992; Pember et al., 1995; Almgren et al., 1997). Similarly to IBMs, in this case also, the transfer of information from the embedded surface to the underlying Cartesian grid needs special treatments to satisfy the conservation laws at the irregularly shaped boundary cells (Cartesian mesh cells cut by the interface). Various cell-cut algorithms have been developed by Forrer and Jeltsch (1998) and Aftosmis et al. (2000), while accuracy assessments have been carried out by Coirier (1994) for the compressible Navier–Stokes equations and by Almgren et al. (1997) for the Poisson equations; full second-order accuracy was obtained considering the cells not intersected by the boundaries.

Finally, an entire class of numerical techniques has been developed with the objective of handling moving interfaces (of various physical natures) on a fixed grid. Two of these are the volume-of-fluid (VOF) and the level-set methods (Hirt and Nichols, 1981; Brackbill et al., 1992; Sussman and Puckett, 2000). The main application of these techniques is the tracking of interfaces between immiscible fluids (Popinet, 2003; Tryggvason et al., 2011) or the propagation of flames in combustion (Pitsch and Duchamp de Lageneste, 2002). Nevertheless, the boundary between a fluid and a solid can also be thought of as an interface and so the aforementioned techniques can be used in a context similar to the IBMs (Lorstadt et al., 2004; Cottet and Maitre, 2004).

1.4 Objectives and Plan

This book aims to provide the reader with the main concepts of the immersed boundary methods with a particular focus on their application to direct numerical simulation of complex flows. In Chapters 2–9 the required tools will be explained and discussed: governing equations, forcing methods, boundary conditions and reconstruction, moving/deforming boundaries and fluid–structure interaction. Some basic concepts of computational geometry will also be given in order to efficiently handle complex three-dimensional geometries in the IB context.

The final chapter (Chapter 10) of this book describes some numerical examples of IBMs applied to flows of increasing complexity and suggestions for further developments are also provided.

Some of the applications can also be viewed as a tutorial to get familiar with the use of the IB flow solver available with this book.

Finally, it is worthwhile pointing out that, although we have tried to give a comprehensive overview of the subject, the reader must be aware that the field is evolving so rapidly that up-to-date material can only be found in the current issues of archival journals.

2

Introduction to Flow Simulations

2.1 Fundamentals of Fluid Dynamics

Although this book focuses mainly on the set of numerical techniques referred to as immersed boundary methods (IBMs), some general preliminary considerations are necessary to ensure self-consistency and to clarify the discussions that follow.

For this reason, in this chapter we will briefly introduce some basic concepts about fluids, the equations governing their dynamics and the issues related to the numerical integration of these equations.

We feel that this information is necessary since immersed boundary techniques must be coupled with numerical algorithms already established for specialized solvers, which must cope with specific flow features.

It is worthwhile mentioning that each of the points mentioned below is a topic deserving a textbook in its own: Possible references are Anderson (1995), Fletcher (1996), Chung (2002) and Ferziger and Peric (2002), just to mention a few popular titles among many others.

2.1.1 Definition of Fluid

Fluid dynamics is the branch of continuum mechanics dealing with *fluids* that can be gases, liquids or even "solids" depending on the response of the material to a specific external load.

In order to make a simple initial distinction, let us consider a parallelepiped of an unknown material to which a tangential stress is applied in the time window $t_1 \leq t \leq t_2$ as shown in Figure 2.1a. Among several possibilities, two distinctive behaviors are observed: In the first, the material undergoes a finite angular deformation and when the stress is removed it recovers its initial shape (Figures 2.1b,d,f). In the second case, the parallelepiped shows a constantly

16

increasing angular deformation and does not return to its original shape after the load is removed (Figures 2.1c,e,g). The former behavior is that of elastic solids, while the latter materials are referred to as *fluids*. Regardless of the particular nature of the matter (solid, liquid or gas), the reason for the second behavior is the absence of permanent links among atoms or molecules, although it is worth mentioning that the details of their microscopic interactions produce a wide variety of different fluids whose characterization is the subject of rheology.

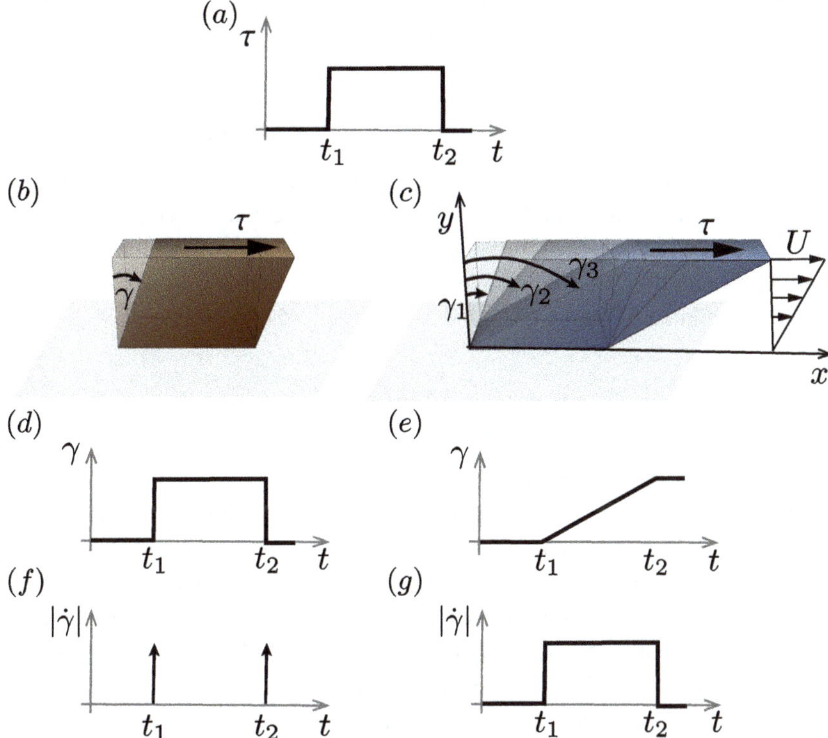

Figure 2.1 Time window of the tangential stress application (a); deformation of an elastic solid (b) and of a fluid (c); angular deformation for the solid (d) and for the fluid (e); angular deformation rate for the solid (f) and for the fluid (g).

If the angular deformation increases at a constant rate, as in Figure 2.1e, its time derivative $\dot{\gamma}$, the deformation rate, is constant in the time interval $t_1 \leq t \leq t_2$ and it can be related to the local velocity gradient through $\dot{\gamma} = \mathrm{d}U/\mathrm{d}y$ (Figure 2.1c).

For many fluids, but not for all of them, the experimental evidence suggests that there is a direct proportionality between the angular deformation rate and the tangential stress through the relation

$$\tau = \mu \frac{\mathrm{d}U}{\mathrm{d}y}, \tag{2.1}$$

where μ is the dynamic viscosity, which is solely a property of the fluid and its state. The fluids that obey equation (2.1) are referred to as Newtonian fluids.

The most common fluids, such as water and air, but also all perfect gases and "simple" liquids (alcohols, kerosene, mercury), indeed behave according to equation (2.1), which yields a relation between *local* fluid deformation and viscous stress whose general form will be introduced later.

2.1.2 Reynolds Transport Theorem

A finite continuum system can be thought of as a collection of infinitesimal particles, each one small enough for its properties to be considered homogeneous in it. Let $\mathbf{x}$ be the position of one of them at time t and $b(\mathbf{x}, t)$ a generic local intensive quantity. The conjugate extensive quantity $B(t)$ for the whole system is then computed through

$$B(t) = \int_{V(t)} \rho(\mathbf{x}, t) b(\mathbf{x}, t) \mathrm{d}V, \tag{2.2}$$

where ρ is the local value of the material density.

If the system is composed always by the same particles it is referred to as a *material volume* $V(t)$ and its boundary is referred to as a *material surface* $S(t)$. An immediate consequence is that each point of $S(t)$ must move in space with the same local velocity as the continuum since otherwise either external particles would enter the system or internal ones would leave it, which is against the initial hypothesis. This implies that, in general, our system will move and deform in time and therefore, according to equation (2.2), B can vary in time not only because of b and ρ but also because of V.

In order to derive the governing equations for a continuum system, it is useful to compute the rate of change of $B(t)$ as

$$\frac{\mathrm{d}B(t)}{\mathrm{d}t} = \frac{\mathrm{d}}{\mathrm{d}t} \int_{V(t)} \rho(\mathbf{x}, t) b(\mathbf{x}, t) \mathrm{d}V, \tag{2.3}$$

which can also be written as

$$\frac{\mathrm{d}B(t)}{\mathrm{d}t} = \lim_{\Delta t \to 0} \frac{\int_{V(t+\Delta t)} \rho b \, |_{t+\Delta t} \, \mathrm{d}V - \int_{V(t)} \rho b \, |_{t} \, \mathrm{d}V}{\Delta t}. \tag{2.4}$$

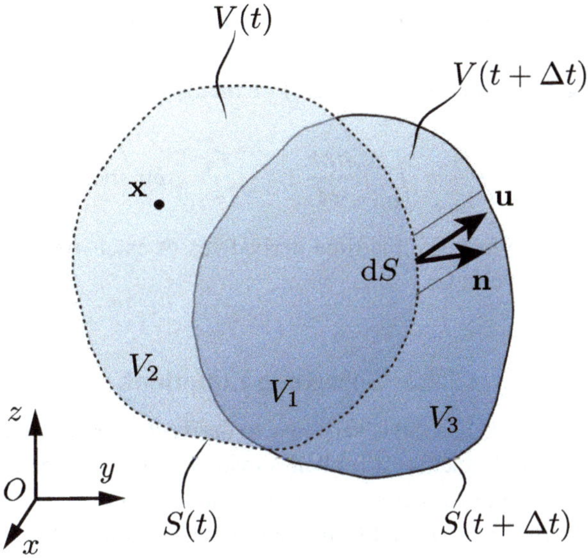

Figure 2.2 Deformation of a material volume in time.

Following the sketch of Figure 2.2 we have $V(t) = V_1 + V_2$ and $V(t + \Delta t) = V_1 + V_3$ so that equation (2.4) becomes

$$\frac{\mathrm{d}B(t)}{\mathrm{d}t} = \lim_{\Delta t \to 0} \frac{\int_{V_1} \rho b \mid_{t+\Delta t} \mathrm{d}V + \int_{V_3} \rho b \mid_{t+\Delta t} \mathrm{d}V - \int_{V_1} \rho b \mid_t \mathrm{d}V - \int_{V_2} \rho b \mid_t \mathrm{d}V}{\Delta t}.$$

(2.5)

Upon noticing that the first and third terms of equation (2.5) are both computed on V_1 we can lump them together to obtain:

$$\lim_{\Delta t \to 0} \frac{\int_{V_1} \rho b \mid_{t+\Delta t} \mathrm{d}V - \int_{V_1} \rho b \mid_t \mathrm{d}V}{\Delta t} = \lim_{\Delta t \to 0} \int_{V_1} \frac{\rho b \mid_{t+\Delta t} - \rho b \mid_t}{\Delta t} \mathrm{d}V$$

$$= \int_{V(t)} \frac{\partial \rho b}{\partial t} \mathrm{d}V.$$

(2.6)

The second and fourth terms of equation (2.5) are instead transformed into surface integrals using the relation $\mathrm{d}V = \mathbf{u} \cdot \mathbf{n} \mathrm{d}S \Delta t$ thus yielding

$$\lim_{\Delta t \to 0} \frac{\int_{V_3} \rho b \mid_{t+\Delta t} \mathrm{d}V - \int_{V_2} \rho b \mid_t \mathrm{d}V}{\Delta t}$$

(2.7)

$$= \lim_{\Delta t \to 0} \left[\int_{S_{13}} \rho b \mathbf{u} \cdot \mathbf{n} \mid_{t+\Delta t} \mathrm{d}S + \int_{S_{12}} \rho b \mathbf{u} \cdot \mathbf{n} \mid_t \mathrm{d}S \right] = \int_{S(t)} \rho b \mathbf{u} \cdot \mathbf{n} \mathrm{d}S,$$

with S_{13} and S_{12} the interfaces between V_1/V_3 and V_1/V_2, respectively.

Note that on S_{12} we have used $\mathrm{d}V = -\mathbf{u} \cdot \mathbf{n}\mathrm{d}S\Delta t$ since the angle between $\mathbf{u}$ and $\mathbf{n}$ is obtuse.

Putting together equations (2.5)–(2.7) we finally have the Reynolds transport theorem:

$$\frac{\mathrm{d}B(t)}{\mathrm{d}t} = \int_{V(t)} \frac{\partial \rho b}{\partial t}\mathrm{d}V + \int_{S(t)} \rho b\mathbf{u} \cdot \mathbf{n}\mathrm{d}S, \qquad (2.8)$$

which allows us to evaluate the time derivatives of extensive quantities computed on material volumes.

2.1.3 Governing Equations

With equation (2.8) at hand, it is very easy to derive the governing equations for a material volume. In fact, as it is always made by the same material particles its mass $M(t)$ must remain constant in time: Setting $b(\mathbf{x}, t) = 1$ we have from equation (2.2) $B(t) = M(t)$ and from the Reynolds transport theorem (2.8)

$$\frac{\mathrm{d}M(t)}{\mathrm{d}t} = \int_{V(t)} \frac{\partial \rho}{\partial t}\mathrm{d}V + \int_{S(t)} \rho\mathbf{u} \cdot \mathbf{n}\mathrm{d}S = 0, \qquad (2.9)$$

or, assuming the hypotheses of the divergence theorem apply,

$$\int_{V(t)} \left[\frac{\partial \rho}{\partial t} + \nabla \cdot (\rho\mathbf{u}) \right] \mathrm{d}V = 0. \qquad (2.10)$$

Since the above equation must be valid for *any* material volume $V(t)$, the localization lemma allows us to write the corresponding differential equation

$$\frac{\partial \rho}{\partial t} + \nabla \cdot (\rho\mathbf{u}) = 0, \qquad (2.11)$$

which enforces the mass conservation locally.

We can now put in the definition (2.2) $b(\mathbf{x}, t) = \mathbf{u}(\mathbf{x}, t)$ to compute the momentum of the system $B(t) = \mathbf{Q}(t)$ whose rate of change equals the resultant of the applied forces $\mathbf{F}$, according to Newton's second law of motion. From equation (2.8) we obtain

$$\frac{\mathrm{d}\mathbf{Q}(t)}{\mathrm{d}t} = \int_{V(t)} \frac{\partial \rho\mathbf{u}}{\partial t}\mathrm{d}V + \int_{S(t)} \rho\mathbf{u}\mathbf{u} \cdot \mathbf{n}\mathrm{d}S = \mathbf{F}, \qquad (2.12)$$

and using standard expressions and the Cauchy stress theorem

$$\mathbf{F} = -\int_{S(t)} p\mathbf{n}\mathrm{d}S + \int_{S(t)} \boldsymbol{\tau} \cdot \mathbf{n}\mathrm{d}S + \int_{V(t)} \rho\mathbf{f}\mathrm{d}V, \qquad (2.13)$$

for pressure, viscous and volume forces (p, $\boldsymbol{\tau}$ and $\mathbf{f}$ are, respectively, pressure, deviatoric viscous stress tensor and specific volume force vector), we can write

$$\int_{V(t)} \frac{\partial \rho \mathbf{u}}{\partial t} dV + \int_{S(t)} \rho \mathbf{u} \mathbf{u} \cdot \mathbf{n} dS = - \int_{S(t)} p \mathbf{n} dS + \int_{S(t)} \boldsymbol{\tau} \cdot \mathbf{n} dS + \int_{V(t)} \rho \mathbf{f} dV. \tag{2.14}$$

Similarly to the mass conservation, we can use the divergence theorem and localization lemma to obtain the balance of momentum in differential form

$$\frac{\partial \rho \mathbf{u}}{\partial t} + \nabla \cdot (\rho \mathbf{u} \mathbf{u}) = -\nabla p + \nabla \cdot \boldsymbol{\tau} + \rho \mathbf{f}, \tag{2.15}$$

which could be solved once the tensor $\boldsymbol{\tau}$ is given an explicit expression.

We finally compute the total energy of the system $E(t)$ from (2.2) by posing $b(\mathbf{x}, t) = \mathcal{E}(\mathbf{x}, t)$ with $\mathcal{E} = u^2/2 + e$ the specific energy, which is the sum of kinetic and internal energy. According to the first principle of thermodynamics, the rate of change of E must equal the sum of mechanical and thermal power applied to the system or

$$\frac{dE(t)}{dt} = - \int_{S(t)} p \mathbf{u} \cdot \mathbf{n} dS + \int_{S(t)} (\boldsymbol{\tau} \cdot \mathbf{u}) \cdot \mathbf{n} dS + \int_{V(t)} \rho \mathbf{f} \cdot \mathbf{u} dV + \tag{2.16}$$
$$\int_{V(t)} \rho \dot{q} dV + \int_{S(t)} \lambda \nabla T \cdot \mathbf{n} dS,$$

where the first three terms on the right-hand side are the mechanical power associated with the forces of equation (2.13), while the last two terms are the thermal power due to volumetric heating ($\dot{q}$) and surface heat flux (λ is the thermal conductivity and T is the temperature field).

Using the Reynolds transport theorem (2.8) together with the divergence and localization theorems, the first principle of thermodynamics becomes

$$\frac{\partial \rho \mathcal{E}}{\partial t} + \nabla \cdot (\rho \mathbf{u} \mathcal{E}) = -\nabla \cdot (p \mathbf{u}) + \nabla \cdot (\boldsymbol{\tau} \cdot \mathbf{u}) + \rho \mathbf{f} \cdot \mathbf{u} + \rho \dot{q} + \nabla \cdot (\lambda \nabla T), \tag{2.17}$$

which is the equation of energy conservation in differential form.

2.1.4 Navier–Stokes Equations for Incompressible Flows

Equations (2.11), (2.15) and (2.17) must be completed with a constitutive relation for the tensor $\boldsymbol{\tau}$, a definition for the internal energy e and an equation of state $f(\rho, p, T) = 0$ relating density, pressure and temperature (for example that for ideal gases).

It can be shown that the general constitutive relation for Newtonian fluids reads

$$\boldsymbol{\tau} = 2\mu \mathbf{S} - \frac{2}{3}\mu (\nabla \cdot \mathbf{u}) \mathbf{I}, \tag{2.18}$$

where $\mathbf{S} = (\nabla\mathbf{u} + (\nabla\mathbf{u})^T)/2$ is the rate of strain tensor, which is the symmetric part of the velocity gradient tensor. Equation (2.18) reduces to equation (2.1) for the simple deformation of the configuration in Figure 2.1. If the internal energy depends only on temperature (for example $e = CT$, with C the specific heat), the system of equations (2.11), (2.15) and (2.17) with the equation of state is a closed system of six equations in the six unknowns ρ, p, T and the three velocity components of $\mathbf{u}$, which can be solved once the appropriate initial and boundary conditions are defined.

These governing equations can be simplified if density ρ can be considered constant since velocity $\mathbf{u}$ and pressure p are four unknowns (plus the equation for temperature, if needed), and equations (2.11) and (2.15) are enough to close the system. Within this hypothesis, the equations reduce to

$$\nabla \cdot \mathbf{u} = 0, \tag{2.19}$$

$$\frac{\partial \mathbf{u}}{\partial t} + \nabla \cdot (\mathbf{uu}) = -\frac{\nabla p}{\rho} + \nu\nabla^2\mathbf{u} + \mathbf{f}, \tag{2.20}$$

with $\nu = \mu/\rho$ the kinematic viscosity. These are referred to as Navier–Stokes equations for an incompressible flow and they will be the model mostly used within this book.

It is worth mentioning that in order for equation (2.11) to become (2.19), the condition $\rho = const.$ is not necessary and it can be relaxed to $D\rho/Dt = \partial\rho/\partial t + \mathbf{u}\cdot\nabla\rho = 0$, which requires the density of each infinitesimal fluid particle to remain constant along its trajectory. In this case, however, an additional equation is needed to determine the density field.

Also, equation (2.17) can be simplified using incompressibility, neglecting the heat produced by friction ($\nabla \cdot (\tau \cdot \mathbf{u})$) and subtracting equation (2.20) multiplied by $\mathbf{u}$ to obtain

$$\frac{\partial T}{\partial t} + \nabla \cdot (\mathbf{u}T) = \dot{q} + \kappa\nabla^2 T, \tag{2.21}$$

where $\kappa = \lambda/(\rho C)$ is the thermal diffusivity. Equation (2.21) governs the convection and diffusion of the temperature field T and a similar equation would hold true for other scalar fields, provided the volumetric source term $\dot{q}$, if any, and the diffusivity κ were replaced by those for the scalar.

2.2 Numerical Solution of Equations

Even if systems (2.11), (2.15), (2.17) or (2.19), (2.20) are closed, the equations are nonlinear and coupled, thus yielding, in general, intractable mathematical problems. For this reason, their analytical solution can be obtained only for very special cases in which symmetries or simplifying assumptions are imposed. Usually, the resulting solutions have limited applicability, although they are often used as benchmarks for approximated methods or as illustrative academic examples.

A standard way to tackle partial differential equations when analytical solutions are not available is to replace the continuum relations by some *discretized* approximations which can be solved using numerical procedures. Ironically, we can think of this circumstance as a revenge of Nature; since we have ignored the discrete nature of matter, made by atoms and molecules, assuming a continuum model, the resulting governing equations are so complex that they can be solved only using discrete methods!

As an example, we consider a simple one-dimensional partial differential equation (PDE) describing the evolution of a quantity Q within the domain $a \leq x \leq b$:

$$\frac{\partial Q}{\partial t} + c \frac{\partial Q}{\partial x} = 0, \tag{2.22}$$

with c being constant and $Q(x, t)$ unknown. The solution of equation (2.22) is defined once initial conditions $Q(x, 0) = Q_0(x)$ and boundary conditions $Q(a, t) = Q_a(t)$ and $Q(b, t) = Q_b(t)$ are specified (see Figure 2.3).

Even if we know that equation (2.22) admits the solution $Q(x, t) = Q_0(x - ct)$ (with $Q_a(t) = Q_0(a - ct)$ and $Q_b(t) = Q_0(b - ct)$) we pretend not to know it and discretize the spatial domain by M nodes so that $x_i = a + (b - a)(i - 1)/(M - 1)$ for $i = 1, 2, \ldots, M$. We do the same for the time interval $[0, \overline{T}]$ using N points $t^n = \overline{T}(n - 1)/(N - 1)$ for $n = 1, 2, \ldots, N$ and denote $Q_i^n = Q(x_i, t^n)$. Note that the above analytical solution might entail the knowledge of $Q_0(x)$ beyond the domain of interest $[a, b]$, in order to enforce boundary conditions.

We can now approximate the derivatives of equation (2.22) using the relations

$$\frac{\partial Q}{\partial t} \approx \frac{Q_i^{n+1} - Q_i^n}{\Delta t} \quad \text{and} \quad \frac{\partial Q}{\partial x} \approx \frac{Q_i^n - Q_{i-1}^n}{\Delta x} \tag{2.23}$$

where $\Delta t = t^{n+1} - t^n$ and $\Delta x = x_i - x_{i-1}$.

Plugging the expressions (2.23) into (2.22) and solving for Q_i^{n+1} one obtains

$$Q_i^{n+1} \approx Q_i^n - \frac{c \Delta t}{\Delta x} (Q_i^n - Q_{i-1}^n) \quad \text{for} \quad i = 2, 3, \ldots, M-1 \quad \text{and} \quad n = 2, 3, \ldots, N. \tag{2.24}$$

Note that $Q_i^1 = Q_0(x_i)$, $Q_1^n = Q_a(t^n)$ and $Q_M^n = Q_b(t^n)$ are all known from initial and boundary conditions, therefore equation (2.24) yields a recursive relation that allows the calculation of the solution at any point x_i and time t^n. The aforementioned numerical solution is said to be a *consistent* ap-

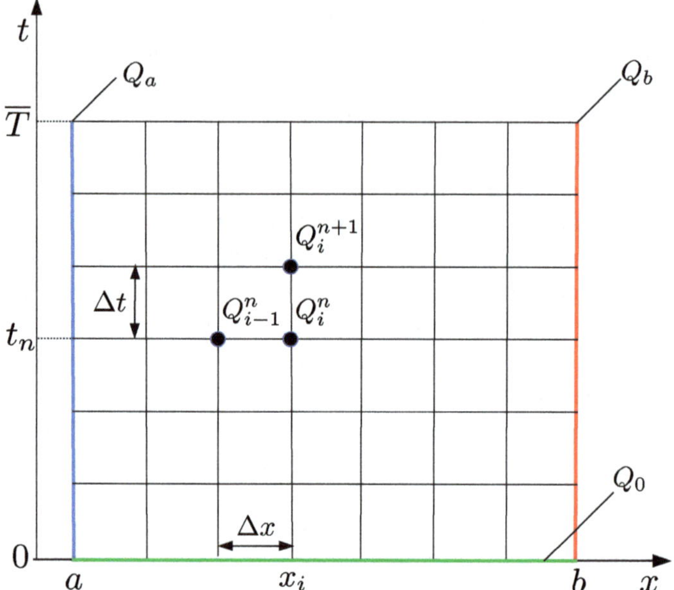

Figure 2.3 Sketch of space and time discretization for the solution of equation (2.22).

proximation of equation (2.22) since as $N, M \to \infty$ (or $\Delta t, \Delta x \to 0$) the derivatives (2.23) become exact and so does the solution (2.24). Unfortunately, calculations can only be performed for finite values of N, M and a simple Taylor-series expansion shows the solution (2.24) to be affected by truncation errors, which determine the accuracy of the scheme; in this case, the scheme is first-order accurate both in time $O(\Delta t)$ and space $O(\Delta x)$. It should also be noted that, since the spatial derivative is evaluated explicitly at the time level n, the integration of (2.24) is conditionally stable, meaning that the numerical error could grow exponentially if the time step exceeds the threshold $\Delta t \leq \Delta x/c$.[1]

We will discuss in more detail the accuracy and stability of numerical schemes throughout the following chapters; here it suffices to mention that the main purpose of the discretization is to transform a continuum differential equation into a system of algebraic equations whose solution can be efficiently obtained by numerical methods and fast computers.

---

[1] On the other hand, a centered finite-difference scheme in space $\partial Q / \partial x \approx (Q_{i+1}^n - Q_{i-1}^n)/(2\Delta x)$, which is second-order accurate, would yield an unstable scheme for any Δt with the same explicit temporal scheme.

2.3 Features of a Numerical Method

The simple previous example has been used to introduce the main aspects of a numerical method that consists of several interconnected components. The first is the definition of the computational domain and its discretization into elements within which the unknowns vary in a prescribed (linear, quadratic, . . .) way. The second is a collection of suitable expressions which approximate all terms of the continuum model with the desired accuracy, that is with an error that decreases with some positive power of the discretization steps (Δx^p, Δt^q). Finally, once the original system of continuum equations has been transformed into a set of algebraic equations, a suitable solution procedure must be devised that provides the solution within a reasonable computational effort.

A fundamental feature of a numerical method is its convergence or, in other words, the property of the discrete solution to approach at some rate the exact counterpart as the space–time domain is tessellated with a finer mesh. For a linear equation like (2.22), we can rely on the Lax equivalence theorem (Lax and Richtmyer, 1956) stating that: *for a linear initial valued problem and a consistent discretization scheme, stability is the necessary and sufficient condition for convergence.*

Unfortunately, a similar theorem does not exist for general nonlinear PDEs, and we can use it as a guideline, at most. On the other hand, for problems usually encountered in fluid dynamics, even the exact solution is not known; therefore, we can only verify that as the discretization is refined, the computed solution becomes grid independent and that some global properties, like total mass or energy conservation (the latter in the inviscid limit), are respected.

The absence of rigorous criteria for the determination of space and time discretization makes choices difficult since the desire to have a small numerical error would suggest adopting the smallest possible Δx and Δt. However, the concurrent growth of N and M also increases the operation count of the numerical method (at a rate faster than linear), possibly resulting in an infeasible numerical simulation.

It is worth mentioning that, up to now, we have discussed only mathematical and numerical aspects of our equations without taking into account the physics of the system they model. The run parameters of a numerical simulation, however, can not leave aside the latter aspect and indeed specific features of the flow physics determine the refinement of the discretization and the numerical model.

To elucidate this point, we consider the flow around a square obstacle with edges of size D as shown in Figure 2.4a. The solution is obtained by integrating numerically equations (2.19) and (2.20) using different values of viscosity, and instantaneous snapshots of the flow are reported in Figure 2.4. As fluid particles approach the body, they have to deviate from the rectilinear trajectory and accelerate in order to maintain the same flow rate through a reduced cross section. On the other hand, the fluid sweeping the solid surface slows down owing to friction. As a consequence, the flow, initially uniform, develops a velocity gradient and, in turn, viscous stresses (see equation (2.18)). Although these features are common to all flow configurations, it is also evident that depending on the value of viscosity the flow develops dynamics with different levels of complexity.

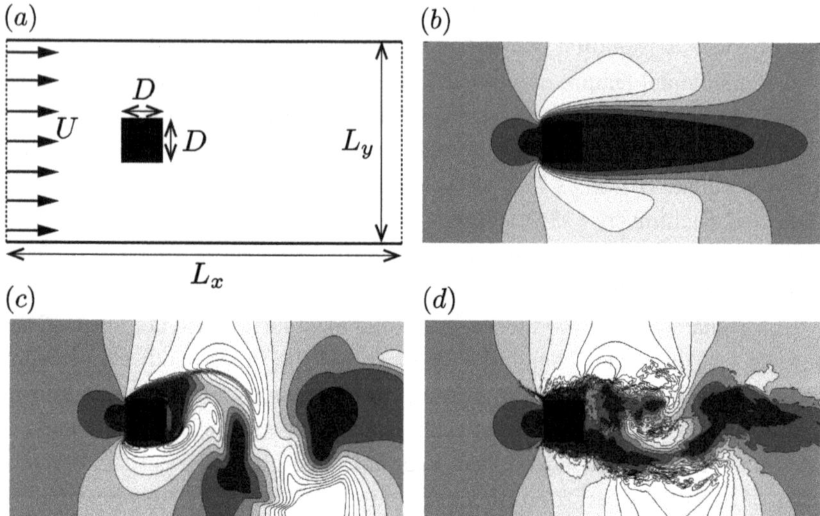

Figure 2.4 Flow around a square cylinder for different values of the flow viscosity: (a) setup of the problem; (b) instantaneous snapshot of the velocity magnitude for $v = UD/50$; (c) the same as (b) but for $v = UD/500$; (d) the same as (b) but for $v = UD/5000$. The grayscale ranges from $u = 2U$ (white) to $u = 0.2U$ (darkest gray). The simulation is performed for a three-dimensional z-periodic configuration with a domain dimension perpendicular to the page equal to D; in these panels only the middle x–y plane is shown.

In particular, let ρ be the fluid density and $v = \mu/\rho$ the kinematic viscosity; for the highest value $v = UD/50$, Figure 2.4b shows a smooth, symmetric and steady flow field as would be expected from a configuration with a symmetry plane, constant uniform inflow and stationary boundary conditions.

As viscosity is decreased by one order of magnitude ($v = UD/500$, Figure 2.4c), however, the same flow develops a completely different dynamics with slow and fast fluid patches which are alternatively shed from the upper and lower rear corners, thus producing a wavy wake downstream. Another evident feature is the decreased flow smoothness and the local concentration of velocity isolines, indicating stronger velocity gradients.

All these phenomena are exacerbated when viscosity is reduced further, as shown in Figure 2.4d, for $v = UD/5000$, although a closer inspection of the results reveals also some wiggling of the velocity isolines at a fixed wavelength that was not observed in the previous cases. In order to understand whether these small-scale oscillations depend on physical or numerical instabilities, we could repeat the same calculation using a much finer spatial discretization and find that most of the wiggles would fade away, even if a further viscosity reduction would produce unphysical wiggles again.

From this simple example, we have learned some important facts about the motion of fluids and their numerical simulation. The first is that, when decreasing viscosity,[2] even a problem with stationary and symmetric boundary conditions can develop an unsteady flow, whose instantaneous realizations do not comply with any symmetry.

The second observation is that for a given flow, the smaller the viscosity, the steeper the velocity gradients and the tinier the size of the produced flow structures. Finally, we have realized that any discretization can capture features only up to a given length scale, below which the numerical solution develops localized oscillations whose origin is not physical.

The latter finding is a known phenomenon occurring when a continuum signal is discretized using a too-coarse sampling (mesh) and the problem has been theoretically addressed by the Nyquist–Shannon sampling theorem (Nyquist, 1928; Shannon, 1948): If a continuous signal contains no wavelength components shorter than $\lambda_{\min}$, then it can be completely determined by uniform samples taken at a spacing $\Delta_{\max} = \lambda_{\min}/2$. The implications of this theorem for our application are that in order to properly represent a flow structure of size λ our mesh has to be fine enough to fit at least two elements in it; if the resolution is not adequate, all the wavelengths smaller than 2Δ are misrepresented as bigger wavelengths (aliasing) and this explains how the unphysical oscillations observed in Figure 2.4d are generated.

According to the Nyquist–Shannon sampling theorem, and considering the results of Figure 2.4, it is then clear that any discretization can adequately describe the flow only up to a threshold viscosity below which the numerical solution becomes inaccurate. To understand, however, how small flow structures

---

[2]Indeed, it is not viscosity alone that matters but a nondimensional parameter, the Reynolds number $Re = UD/v$, which for high enough values triggers flow transition to chaotic dynamics and eventually to turbulence.

can be produced by large ones, even if the forcing is smooth and regular, we have to look in more detail at the structure of the Navier–Stokes equations.

2.4 Energy Cascade and Turbulence

In Section 2.2 we used the simple equation (2.22) to illustrate the main issues of the spatial discretization size and its interaction with the time step size. Here we resort to another model equation to show the very different role of convective and viscous terms of the Navier–Stokes equation (2.20).

Upon subtracting equation (2.19) multiplied by $\mathbf{u}$ from equation (2.20) we obtain

$$\frac{\partial \mathbf{u}}{\partial t} + \mathbf{u} \cdot \nabla \mathbf{u} = -\frac{\nabla p}{\rho} + \nu \nabla^2 \mathbf{u} + \mathbf{f}, \tag{2.25}$$

in which the second terms on the left- and right-hand side are, respectively, the convective and viscous terms.

Even within the simplifying assumption of constant density ($\rho = const.$), equation (2.25) still has four unknowns ($\mathbf{u}$ and p) and must therefore be completed with the mass conservation (2.11) which, with the condition $\rho = const$, reduces to $\nabla \cdot \mathbf{u} = 0$. The latter and equation (2.25) constitute a closed system of coupled PDEs in four unknowns which, however, is difficult to treat analytically.

A simpler one-dimensional surrogate of (2.25), often used to mimic the complete model, is the Burgers equation

$$\frac{\partial u}{\partial t} + u \frac{\partial u}{\partial x} = \nu \frac{\partial^2 u}{\partial x^2}, \tag{2.26}$$

which is still a PDE, with the same structure of convective and viscous terms as equation (2.25), although it contains only one scalar unknown which depends on a single spatial direction ($u(x,t)$).

A straightforward way to assess the different effects of its terms is to assume that the unknown can be expressed as

$$u(x, t) = \sum_{k=1}^{\infty} \widetilde{u}_k(t) \sin(kx), \qquad 0 \le x < 2\pi, \tag{2.27}$$

where the integer k is the wavenumber and is equal to the number of waves that can be fit within the interval $[0, 2\pi]$. Hence, the larger k the smaller the flow structure it represents.

In expression (2.27) we have assumed that our solution always has zero mean and that it is periodic in space over a length 2π. This can be done because the features we want to highlight do not depend on these simplifying hypotheses, even if it can be shown that the same conclusions hold true for the general case.

The various terms of the Burgers equation can be computed analytically from the expansion (2.27) and substituted back into (2.26); relying on the orthogonality property of the sine functions, one finally obtains

$$\frac{\mathrm{d}\widetilde{u}_k(t)}{\mathrm{d}t} + \sum_{m=1}^{\infty} \frac{m}{2} \left(\widetilde{u}_m \widetilde{u}_{k-m} + \widetilde{u}_m \widetilde{u}_{k+m} - \widetilde{u}_m \widetilde{u}_{m-k} \right) = -\nu k^2 \widetilde{u}_k, \quad k = 1, 2, \ldots, \infty,$$

(2.28)

which governs the time evolution of the individual mode amplitude $\widetilde{u}_k$.

Equation (2.28) immediately shows that each viscous term can change only its own magnitude; therefore, there is no way for viscosity to create modes that did not already exist. In contrast, the nonlinear terms (the convective part) indicate that there are infinite possibilities for two modes to alter the dynamics of a third one, and this includes the possibility of creating a mode that was not present initially. In other words, even if the initial conditions are such as to restrict the motion only of the modes within a narrow band, the nonlinear terms can transfer their energy to smaller and larger scales to generate a continuum spectrum.

Concerning energy, other important insights can be obtained by computing it in terms of modes:

$$E(t) = \frac{1}{2} \int_0^{2\pi} u^2 \mathrm{d}x = \frac{1}{2} \sum_{l,m=1}^{\infty} \widetilde{u}_l \widetilde{u}_m \int_0^{2\pi} \sin(lx)\sin(mx)\mathrm{d}x \qquad (2.29)$$

$$= \frac{\pi}{2} \sum_{l,m=1}^{\infty} \widetilde{u}_l \widetilde{u}_m \delta_{l,m} = \frac{\pi}{2} \sum_{k=1}^{\infty} \widetilde{u}_k^2,$$

where $\delta_{l,m}$ is the Kronecker delta-function equal to one for $l = m$ and zero otherwise.

On the other hand, equation (2.28) multiplied by $\widetilde{u}_k$ and summed over the modes yields

$$\frac{1}{2} \frac{\mathrm{d}}{\mathrm{d}t} \sum_{k=1}^{\infty} \widetilde{u}_k^2 + \sum_{k,m=1}^{\infty} \widetilde{u}_k m \left(\widetilde{u}_m \widetilde{u}_{k-m} + \widetilde{u}_m \widetilde{u}_{k+m} - \widetilde{u}_m \widetilde{u}_{m-k} \right) = -\nu \sum_{k=1}^{\infty} k^2 \widetilde{u}_k^2,$$

(2.30)

whose second term on the left-hand side is identically zero, since for each term of the summation there is an opposite counterpart (the reader can check it directly by limiting the summation to some small value, say 3 or 4, rather than ∞). The previous expression can thus be recast in

$$\frac{\mathrm{d}E(t)}{\mathrm{d}t} = -\pi\nu \sum_{k=1}^{\infty} k^2 \tilde{u}_k^2, \tag{2.31}$$

which balances the rate of change of the total energy with the dissipation rate.

Comparing equations (2.28) and (2.31), we can conclude that although the nonlinear term distributes energy among different scales, it is conservative and can not change the total energy content of the system. In contrast, the viscous term operates only on its own scale, it is always dissipative and it is more effective the smaller the structure is (or the bigger is its wavenumber k).

Further analysis of equation (2.28) reveals that, starting from structures of given sizes, the nonlinear terms create either smaller and bigger eddies, even if, with dissipation more effective for larger k, the overall result is a net energy transfer from larger towards smaller scales. This forward flux is countered only by viscosity, which operates a cut at some wavenumber and prevents even smaller structures from being created.

These observations allow us to interpret the results of Figure 2.4 in the sense that a decreasing viscosity allows the flow to develop increasingly smaller scales and, if the viscosity is not big enough, they become too small for the underlying mesh to correctly describe them. An additional effect of the energy cascade is the growing number of eddies in the flow that increase the degrees of freedom and the complexity of its dynamics. Hence, a flow can become unsteady and asymmetric even if the boundary conditions and the forcings are steady and symmetric.

Of course, the aforementioned interpretation is correct provided the dynamics of equation (2.26) is indeed representative of the system (2.19), (2.20). In fact, it can be shown that the latter behaves very much like the simpler Burgers model when the flow evolves in a three-dimensional space. In two dimensions, instead, the existence of additional flow invariants constrains the flow to generate an inverse energy cascade and, in the long term, it tends to develop larger flow structures.

In this short discussion, we have intentionally left out the dynamics of solid boundaries, whose effect is to produce intense shear layers whose thickness diminishes for decreasing viscosity; the issues related to the numerical simulation of these (boundary) layers will be discussed in more detail in Chapter 8.

It should not have escaped the attentive reader that, in the example of Figure 2.4, the kinematic viscosity ν is always expressed in UD units; therefore, the ratio UD/ν must be a nondimensional parameter. Indeed, using U and D, respectively, as velocity and length scales to make equations (2.19) and (2.20) nondimensional, the resulting system turns out to depend only on UD/ν, that is the Reynolds number Re. Theory and experiments confirm that the flow transitions to turbulence provided the Reynolds number exceeds a threshold that, however, depends on the specific problem. Although the details of the theory are left to specialized textbooks, here we wish to stress that the smallest flow scales can be estimated as $\lambda_{min} \approx D/Re^{3/4}$ far from solid boundaries and $\lambda_{min} \approx D/Re^{0.9}$ in the near-wall region, thus decreasing for increasing Re in both cases. For three-dimensional flows, these estimates yield meshes with $O(Re^{9/4})$ and $O(Re^{2.7})$ elements and, since most real applications have large Reynolds numbers (of the order of millions), the resulting numerical problems become intractable.

In order to make the numerical simulation of high Reynolds number flows feasible there are turbulence models that parametrize the effects of all the unresolved scales, smaller than the mesh, on the resolved ones. This is a difficult topic to which abundant literature is devoted, and we will only briefly discuss some aspects when needed for our numerical examples.

2.5 Modeling, Validation and Verification

As we have seen in Sections 2.3 and 2.4, several steps are necessary in order to obtain a numerical solution of a real flow, the former being only an approximate representation of the latter. A preliminary choice is the set of equations to be solved and, while for a single-phase flow of a Newtonian fluid the Navier–Stokes equations are considered a standard reliable model, for complex flows like multiphase, reactive or non-Newtonian fluid flows the initial set of equations might already be uncertain.

Another aspect of the modeling is related to the complicated nature of real applications to be simulated by a brute force approach and, therefore, simplifications must be introduced. One example is turbulence, which requires extremely fine grids and time steps to be tackled directly (Mahesh and Moin, 1998); in practical applications, this is typically not possible and turbulence models are introduced to capture the main flow features on much coarser meshes.

Once the governing equations are defined, they must be completed by initial and boundary conditions that are often only a model or are known with some degree of approximation.

The final system is then discretized by replacing the differential operators with numerical derivatives performed on a domain that is tessellated (in space and time) by small elements.

Finally, there is the solution step which involves numerical algorithms capable of solving the resulting algebraic discretized system of equations within a finite (though large) number of operations and a limited number of decimals used to represent a real number on computers.

Each of the aforementioned steps introduces some approximations or simplifications that are necessary to distinguish between *errors* and *uncertainties*. For example, substituting numerical derivatives to the differential operators of the equations introduces a truncation error that could be estimated by matching the Taylor-series expansion of the continuum operator with the numerical derivative. Round-off errors are instead caused by the finite-precision arithmetics of computers that retain only a prescribed number of decimals in any floating-point operation. Convergence errors might be produced by the solution step, for example, when an iterative procedure is adopted and it is interrupted when the residue is below a fixed threshold.

Uncertainties, on the other hand, are associated with limited knowledge of the problem, thereby yielding an incomplete or inaccurate definition (for example the detailed roughness profile of the surface when imposing boundary conditions). The main difference between errors and uncertainties is that the former are mostly known and unavoidable, while the latter are potentially present but often ignored.

When evaluating the reliability of CFD predictions, either errors and uncertainties have to be considered. A lot of attention has been devoted in the past few decades to the issue of the accuracy of CFD simulations, which has to do with the rate at which some errors decrease as the discretization is refined, while only more recently has the importance of uncertainty been recognized (Bijl et al., 2013).

The assessment of the CFD quality is nowadays performed by *verification* and *validation* studies.

Verification consists of demonstrating that the calculations satisfy certain theoretical expectations such as the energy conservation in the limit of vanishing viscosity, or the constancy of some invariants (when available). Other verifications are performed by repeating the same calculation on successively refined grids and checking the solution convergence or comparing the results obtained by different codes relying on similar numerical techniques.

Validation, on the other hand, aims at establishing the reliability of the CFD simulations in reproducing the intended problem. For example, if experimental measurements are available, a benchmark with the simulation can be performed. This, however, is possible in only a very few cases when combined experimental measurements and numerical simulations are planned in advance. More often, data are available for canonical flows and these problems are used to validate separate aspects of a complex CFD software.

Whenever possible, for the applications presented in this book both verification and validation studies will be shown to assess the reliability of the results.

3

Computational Geometry and Grid Generation

In Chapter 1, we mentioned that a typical CFD simulation starts from the discretization of the fluid volume in the region of interest to generate a mesh, which is the basis of the computational method. This step can become very challenging if the boundaries are geometrically complex, which is always the case for real applications. The first difficulty stems from the need to construct a high-quality surface mesh, conforming to the boundaries (as in Figure 1.1), that becomes the starting point to generate the volume mesh. The relation between the surface discretization and the computational grid is therefore crucial, although the standard computer-aided design (CAD) software is not optimized to generate CFD-compliant surface meshes but rather to describe accurately the object shape within the minimum amount of information.

An extreme demonstration of this concept is given in Figure 3.1a, showing a turbulent flow in a channel with unstable stratification (Blass et al., 2020). In order to properly capture the tiny flow structures, the lower rectangular boundary is discretized by more than eight million square elements that are extruded in the z-direction to build a volume mesh of more than two billion nodes. The surface discretization of Figure 3.1b is so fine that it appears as a solid black surface and only the enlarged inset can show the real mesh. On the other hand, the basic rectangular planar geometry can be simply described by the coordinates of the four vertices and most CAD packages would define it just by two triangles, as shown in Figure 3.1c.

The connection between the geometry definition and surface grid becomes even more critical for complex nonplanar boundaries and considerable efforts have been made to develop tools to convert CAD-generated geometries into CFD-compliant parametrizations (Aftosmis et al., 1999; Sun et al., 2020).

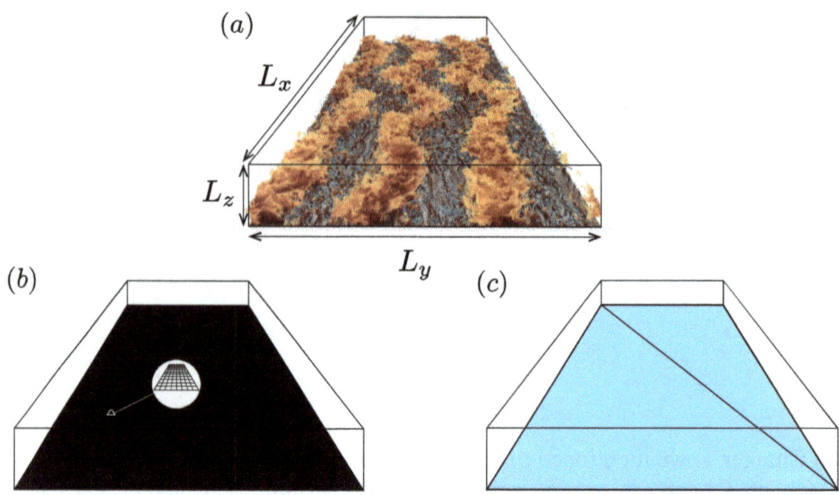

Figure 3.1 (a) Instantaneous snapshot of the turbulent flow structures for a channel flow with unstable stratification (adapted from Blass et al. (2020)). (b) Surface discretization of the domain and inset with an enlarged portion of the grid. (c) Description of the lower rectangular boundary via two triangles obtained by the addition of a diagonal.

Decoupling the geometrical description of the object from the CFD mesh has the twofold advantage of eliminating the need to produce high-quality surface meshes and of generating the computational grid independently of the object geometry: Immersed boundary methods achieve exactly this goal.

3.1 Geometry Definition

Modeling a three-dimensional object is the starting point of many problems of computational science and engineering, such as rapid prototyping, three-dimensional (3D) printing, multibody dynamics and CFD. The accurate geometrical representation of a body is a relevant task of computational geometry and, depending on the specific application, several alternatives are available; for example, primitives like planes, circles, spheres, cubes, etc., can be combined, extruded, stretched or lofted to create more complex shapes. A more flexible alternative consists of patching together parametric surfaces, such as splines or NURBS, that can be deformed locally by moving a set of control points; these surfaces, in turn, can be swept, lofted or rotated to form complex curvilinear bodies. Another classical approach is the polygonal surface modeling that can be achieved either by intersecting the surface with polyhedra (cutting cube

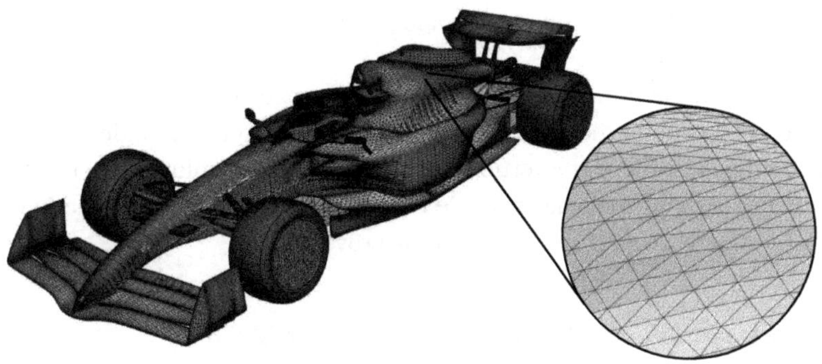

Figure 3.2 Distribution of surface triangles for a racing car model.

method) (Schmidt, 1993) or by a marching procedure (Hartmann, 1998) that, starting from an initial polygon and following given rules, uses the initial edges to propagate new polygons until the whole surface is covered.

Each method has advantages and drawbacks although the polygonal surface modeling is standard in all CAD softwares since any polygon can be further simplified into triangles (theorem: Every polygon having n edges can be decomposed into $n - 2$ triangles by the addition of $n - 3$ diagonals) and the latter have several computational advantages which will be discussed in the following.

Initially, we assume that an object is described by a collection of triangles (either "connected" or "unconnected," respectively, depending on the availability of information about edges and vertices shared among triangles) in space and that they can have an uneven distribution of sizes and skew angles.

An example is given in Figure 3.2 showing a realistic racing car described by a mesh of about 250 000 triangles. Again, it is worthwhile pointing out that this discretization may not be suitable as surface mesh for the generation of body-fitted volume grids if the elements are too big in the low curvature region of the surface and many triangles are highly skewed.

Examples of two widely used standards for the definition of a triangulated object are given next: The first is the "Stereo-LiThography" (STL) format, in which the representation of a surface is a collection of unconnected triangles (Figure 3.2). This description is the standard for the *rapid prototyping* community and most of CAD systems have the ability to export in STL format automatically. This allows the treatment of any complex geometry without the need to generate a surface mesh; the *only* requirement for the object description is that the given surface must be a closed manifold. This is the same restriction enforced by rapid prototyping tools and guarantees that the final objects can be machined (produced).

Apart from the initial header line, the file is made of blocks of seven lines, each one containing the components of the normal to the triangle surface and the coordinates of its three vertices with respect to an absolute reference frame. It is worth noticing that each triangle is defined independently of the others and no information about the connectivity is provided.

```
solid OBJECT
  facet normal -0.039907 0.999200 -0.002476
    outer loop
      vertex -2.905339 -0.448947 -0.980340
      vertex -2.835030 -0.446148 -0.984036
      vertex -2.885250 -0.448157 -0.985734
    endloop
  endfacet
  facet normal -0.997755 0.009863 -0.066224
    outer loop
      vertex -2.206670 -0.531862 -0.927824
      vertex -2.206700 -0.542648 -0.928978
      vertex -2.206809 -0.521854 -0.924225
    endloop
  endfacet
  ...

  ...
endsolid OBJECT
```

The alternative standard "GNU Triangulated Surface" (GTS) makes up for the shortcomings of the STL format by adding connectivity information. In more detail, it describes a group of triangular facets connected by boundary edges. The file starts with three integers, which represent the number of vertices, edges and triangles describing the surface; next, there is a list with the coordinates of each vertex, followed by a list of pairs of integers defining the extreme points of each edge, and finally a list of triplets of integers defining the three edges that bound each triangle. In this way, all essential connectivity is included. In the following example, the geometry is assigned by a cloud of 7 149 points and the lines from 2 to 7 150 assign the coordinates of each one. These points are then connected to form edges whose total number is 21 463 and the corresponding endpoints are given by pairs of integers pointing at the previously defined vertices, which are listed as pairs of integers from line 7 151 to line 28 613 (= 1+7 149+21 463). Finally, triangles are defined by assigning the edges prescribed by triplets of integers (pointing at the edges) in a number of 14 339 and listed from line 28 614 to line 42 952 (= 1 + 7 149 + 21 463 + 14 339). This description gives connectivity information through vertices and edges shared among different triangles, although the local normal is not assigned directly. This information, however, can be easily computed through the coordinates of each triangle vertices provided the integer triplets define them in the correct order for all the normals to point in a consistent direction with respect to the surface.

```
line    |
number  |
1       |    7149 21463 14339
2       |    -0.499595 0.000529 0.004474
3       |    -0.499447 -0.015631 -0.005854
...     |    . . .
7150    |    . . .
7151    |    9 23
7152    |    10 11
7153    |    11 17
...     |    . . .
28613   |    . . .
28614   |    16637 16636 16767
28615   |    5251 4919 4921
...     |    21210 21264 21208
...     |    . . .
42952   |    . . .
```

3.2 Ray Tracing

After having defined the (complex) geometry of an object through the triangulation of its surface, we will now tackle the problem of defining the location of the nodes of a non-body-conformal grid with respect to the body boundaries. Note that the final objective is "just" to separate (*tag*) the computational cells in *solid* (inside the body), *fluid* (outside the body) and *interface* (across the boundary) and, to this aim, any algorithm performing this task could be used.

Here we describe in detail a method based on a ray tracing geometrical algorithm routinely used in computer graphics, because of its scalability and computational efficiency. Namely, a ray, originating from the grid node to be checked, is cast in a random direction and the number of intersections between the ray and the surface is counted; if the total number is even (odd) the

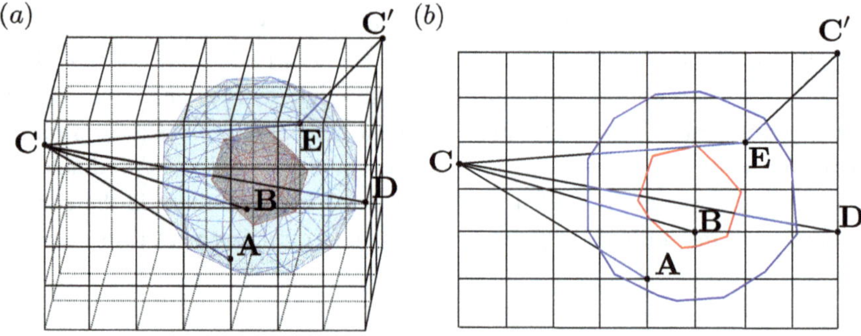

Figure 3.3 (a) Schematic showing a smooth nonconvex body immersed in a Cartesian mesh and crossed by rays cast from the control points C and C'. (b) The same as panel (a) but for a two-dimensional plane cross section.

point is outside (inside) the object. The intersection between a ray (a segment) and the surface (a collection of triangles for STL or GTS surfaces) can be performed using the geometrical algorithms described in O'Rourke (1998) and the essential features are summarized as follows.

In fact, rather than throwing random rays, the easiest and computationally efficient way to perform the tagging of the mesh nodes is by casting a segment between the query point, say A, and a control point C, the latter being definitely outside the body, and counting the number of intersections n of the ray with the surface.

This method is extremely general and it can be applied to any surface provided it is closed (it must enclose an inner volume or, in other words, be "watertight"). The ray tracing works for nonconvex objects and even if the body contains cavities; the only delicate point of the procedure is that the intersections must be counted properly.

Let us consider the example of Figure 3.3a in which a spherical body with a polyhedral cavity is immersed in a Cartesian mesh. For this example, we have considered the points A and E inside the solid, point B inside the inner cavity and therefore outside of the body and point D as an external point. For ease of representation, we have reported, in Figure 3.3b, a two-dimensional cut of Figure 3.3a in which surface triangles become segments. For the inner point A and the outer points B and D the counting of the intersections is straightforward (1, 2 and 4, respectively), but the same is not true for the internal point E whose number of intersections is apparently even ($n = 2$), thus indicating an external point. The contradiction originates from the ray

CE tangent to the inner cavity, which should give a *double* intersection, thus bringing the total number to $n = 3$, which is consistent with the ray-tracing theory. Indeed, since the calculations are performed by a numerical algorithm with floating-point finite-precision arithmetic, deciding whether a segment is just close to a surface or really tangent to it depends on a user-defined tolerance, which makes the final result potentially uncertain. In order to avoid this issue, in those cases in which the distance of the ray from the boundary drops below a given threshold, it is convenient to use a different control point (such as C' in Figure 3.3) from which point E is classified as an internal point with $n = 1$, without any ambiguities.

3.2.1 Segment/Triangle Intersection

Following the previous discussion, it is clear that tagging the nodes of a mesh with respect to an immersed object reduces to computing intersections between rays (segments) and triangles in space, and a naive approach would suggest writing the equation for the line passing through the control and query points and the equation for the plane containing the triangle, which we know through the coordinates of its three vertices. The system of these two equations would give the coordinates of the intersection point (if any); further steps would be then necessary to check if this point belongs to the finite segment, and not just to the infinite line, and if the intersection is indeed inside the triangle rather than just on the infinite plane containing it. It is clear that this procedure requires quite an amount of operations and the tagging can become very expensive for meshes with tens of millions of grid points and surfaces with hundreds of thousands of triangles. An efficient procedure for the parametric computation of the intersections is described in the following, and avoids most of the computational burden and allows fast and efficient tagging on large grids and complex geometries.

Let $\mathbf{Q} = (x_Q, y_Q, z_Q)$ be the generic query point and let $\mathbf{C} = (x_C, y_C, z_C)$ be any control point. The segment $\mathbf{CQ}$ is the ray $\mathbf{r} = (x_Q - x_C, y_Q - y_C, z_Q - z_C)$ and any point belonging to the ray can be spanned by the parameter $u \in [0, 1]$ through the relations

$$x = x_C + u(x_Q - x_C), \quad y = y_C + u(y_Q - y_C) \quad \text{and} \quad z = z_C + u(z_Q - z_C) \quad (3.1)$$

or, in vector form, $\mathbf{x} = \mathbf{C} + u\mathbf{r}$.

In order for the ray $\mathbf{r}$ to intersect a triangle of vertices $\mathbf{A}$, $\mathbf{B}$ and $\mathbf{D}$, it is convenient to verify preliminarily whether the ray intersects the plane containing the triangle. A plane α in space is given by the equation $ax + by + cz = d$ or, equivalently, by the inner product between the vectors $\mathbf{n} = (a, b, c)$ and $\mathbf{x} = (x, y, z)$,

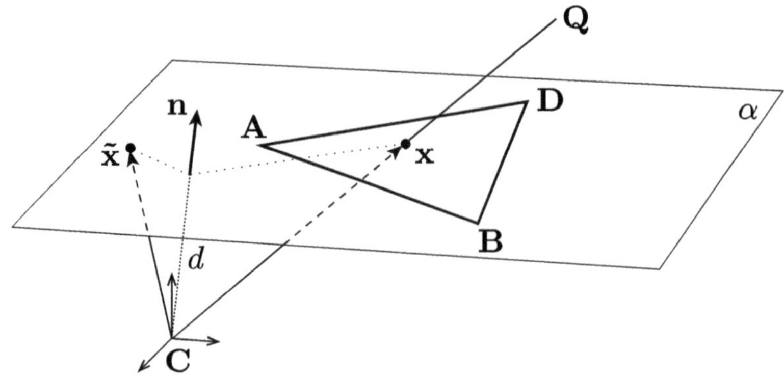

Figure 3.4 Schematic of the segment/triangle intersection in space.

namely $\mathbf{n} \cdot \mathbf{x} = d$. This expression has a simple geometrical interpretation that the vector $\mathbf{x}$, connecting any point on the plane with the origin of the coordinate system ($\mathbf{C}$), has the same projection (d) along the direction $\mathbf{n}$ perpendicular to the plane (Figure 3.4).

It should be noted, however, that given the coordinates of the triangle vertices $\mathbf{A}$, $\mathbf{B}$ and $\mathbf{D}$, we do not know immediately the coefficients a, b, c and d of the plane equation. Of course, the fact that $\mathbf{A}$, $\mathbf{B}$ and $\mathbf{D}$ all belong to the plane could be sufficient to determine, by a linear system, the ratios a/d, b/d and c/d,[1] even if the procedure can be made much simpler and explicit, thus less computationally demanding. In fact, we know that $\mathbf{n}$ is a vector orthogonal to the plane while both the vectors $\mathbf{B} - \mathbf{A}$ and $\mathbf{D} - \mathbf{A}$ lie on the plane. The cross product $(\mathbf{B} - \mathbf{A}) \times (\mathbf{D} - \mathbf{A})$ is therefore parallel to $\mathbf{n}$ from which a, b and c are immediately available up to a constant; the latter can be fixed through the coefficient d obtained by substituting the coordinates of one of the points $\mathbf{A}$, $\mathbf{B}$ and $\mathbf{D}$ in the equation for the plane α.

In more detail, with these positions we have:

$$a = (y_B - y_A)(z_D - z_A) - (z_B - z_A)(y_D - y_A), \tag{3.2}$$

$$b = (x_D - x_A)(z_B - z_A) - (x_B - x_A)(z_D - z_A), \tag{3.3}$$

$$c = (x_B - x_A)(y_D - y_A) - (x_D - x_A)(y_B - y_A), \tag{3.4}$$

$$d = ax_A + by_A + cz_A, \tag{3.5}$$

---

[1] This can be done provided $d \neq 0$. Should be $d = 0$, however, the plane equation contains only a, b and c and the passage of the plane through $\mathbf{A}$, $\mathbf{B}$ and $\mathbf{D}$ is sufficient for the computation of the coefficients.

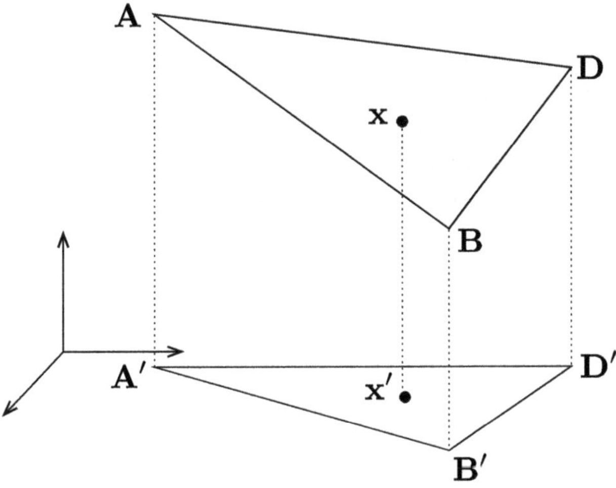

Figure 3.5 Projection of a triangle and its intersection point on a coordinate plane.

which gives the plane coefficients by explicit relations.

With the plane equation at hand, the computation of the intersection between the ray and the plane is immediate upon considering that the condition of a point on the ray $\mathbf{x} = \mathbf{C} + u \cdot \mathbf{r}$ belonging to the plane is $\mathbf{x} \cdot \mathbf{n} = d$ implies

$$(\mathbf{C} + u\mathbf{r}) \cdot \mathbf{n} = d \quad \text{or} \quad u = \frac{d - \mathbf{C} \cdot \mathbf{n}}{\mathbf{r} \cdot \mathbf{n}}, \tag{3.6}$$

which, plugged back into (3.1), yields the coordinates of the intersection point $\mathbf{x}$.

We are now left with the final task of checking whether the intersection $\mathbf{x}$, between the ray and the plane, is inside or outside the triangle. First of all, we note that although both elements are in three dimensions, the intersection problem can in fact be reduced to a two-dimensional problem by introducing a reference frame on the plane of the triangle and considering only the intersection point. Even this rotation, however, is unnecessary when considering that $\mathbf{x}$ is within the triangle only if its projection $\mathbf{x}'$ on one plane is inside the projection of the triangle over the same plane. The projection over the plane orthogonal to the largest component of $\mathbf{n}$ guarantees that degenerate cases like a triangle nearly (or exactly) orthogonal to the plane do not occur (Figure 3.5). In the following, we will denote by the superscript ′ all quantities projected onto the most convenient plane.

As a preliminary step, we consider the problem of computing the area of a triangle. The most usual definition as one half of the product between the base and the height is not practical if the triangle is assigned by the coordinates of the three vertices. An alternative definition, however, can be obtained by the cross product of two vectors. Let $\mathbf{B}' - \mathbf{A}'$ and $\mathbf{D}' - \mathbf{A}'$ be the vectors; their cross product is a third vector orthogonal to the plane containing the original vectors with a magnitude equal to the area of the parallelogram formed by $\mathbf{B}' - \mathbf{A}'$ and $\mathbf{D}' - \mathbf{A}'$. As $\mathbf{A}'$, $\mathbf{B}'$ and $\mathbf{D}'$ are the vertices of the triangle, its area then simply reads

$$\overline{\mathcal{A}}_{A'B'D'} = \frac{1}{2}|(\mathbf{B}' - \mathbf{A}') \times (\mathbf{D}' - \mathbf{A}')|, \tag{3.7}$$

which can be explicitly computed from the coordinates of the vertices. Of course, since in equation (3.7) we take the absolute value of the cross product, the area $\overline{\mathcal{A}}_{A'B'D'}$ will be positive regardless of the relative position of $\mathbf{B}' - \mathbf{A}'$ and $\mathbf{D}' - \mathbf{A}'$. Considering equation (3.7) with its own sign $\mathcal{A}_{A'B'D'} = (\mathbf{B}' - \mathbf{A}') \times (\mathbf{D}' - \mathbf{A}')/2$, however, evidences an interesting property of the area of a triangle, which is positive or negative depending on whether the orientation of the path $A'B'D'$ is counterclockwise or clockwise. This property can be efficiently exploited for the determination of the relative position of a point with respect to a triangle. Consider, in fact, the configuration of Figure 3.6a with the point $\mathbf{x}'$ inside the triangle; the three triangles $A'B'\mathbf{x}'$, $B'D'\mathbf{x}'$ and $D'A'\mathbf{x}'$ thus have all positive areas, respectively, $\mathcal{A}_1$, $\mathcal{A}_2$ and $\mathcal{A}_3$. On the other hand, if the point $\mathbf{x}'$ is outside the triangle, one or two of these triangles will have negative areas (Figures 3.6b,c). If the point $\mathbf{x}'$ is on one of the edges of the triangle ABD one of the areas will be zero (Figure 3.6d). Finally, if two areas are zero the point $\mathbf{x}'$ will be on one of the vertices (Figure 3.6e).

It is worth pointing out that the area $\mathcal{A}_{A'B'D'}$ is positive only if the path $A'B'D'$ is oriented counterclockwise, which is not immediate to verify from the coordinates of the vertices A', B' and D'. However, it is easy to show that, if the sequence $A'B'D'$ is oriented clockwise, the area of the triangle is negative and, if $\mathbf{x}'$ is internal, the triangles $A'B'\mathbf{x}'$, $B'D'\mathbf{x}'$ and $D'A'\mathbf{x}'$ will all have negative areas.

It is then clear that the sign of the area $\mathcal{A}_{A'B'D'}$ is not relevant in itself but only for its agreement or disagreement with the signs of $\mathcal{A}_1$, $\mathcal{A}_2$ and $\mathcal{A}_3$. The criterion can therefore be easily generalized as follows: Given the triangle vertices in an arbitrary sequence, say A', B' and D', if the three subtriangles, $A'B'\mathbf{x}'$, $B'D'\mathbf{x}'$ and $D'A'\mathbf{x}'$, all have areas of the same sign then $\mathbf{x}'$ is inside the triangle $A'B'D'$, while if the sign of one of the areas is different from the other two $\mathbf{x}'$ is outside the triangle. If $\mathcal{A}_1$ is zero, $\mathbf{x}'$ is on the edge $\mathbf{A}'\mathbf{B}'$, while if $\mathcal{A}_1$ and $\mathcal{A}_2$ are both zero, then $\mathbf{x}' \equiv \mathbf{B}'$. All other possibilities can be deduced by induction.

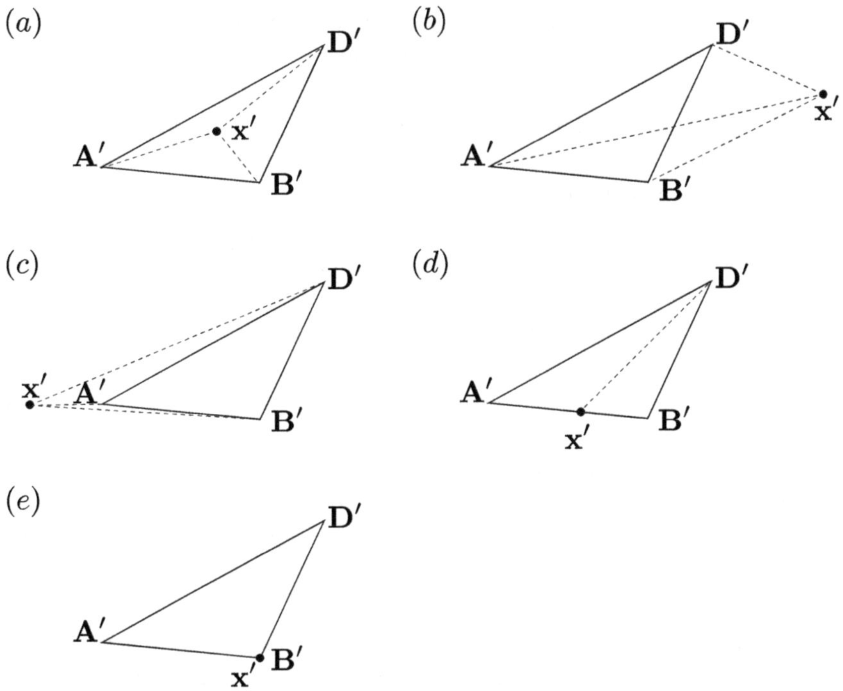

Figure 3.6 Different configurations for the relative position of a point and a triangle.

3.3 Bounding-Box Check

All the above ingredients can be used to set up a procedure for the automatic tagging of the nodes in a computational grid with respect to a generic immersed object.

Usually, grid points and triangles are assigned by a set of coordinates, and although additional information (such as mesh connectivity, block partitioning or surface physical characterization) might be available, this is not necessary for the tagging step. In fact, as shown in Section 3.2, classifying any grid point requires only the counting of valid intersections, which can be obtained through equations (3.6) and (3.7) just using the coordinates of nodes and triangle vertices.

It is worth mentioning, however, that despite the efficient procedure described in Section 3.2, every point must be checked against every triangle. Therefore, for a mesh of N nodes and an object described by M triangles a total number of $N \times M$ interactions must be computed. Typical values of N are of order

10^7–10^8 or more elements, while for realistic complex objects $M \approx 10^5$ is easily achieved, thus implying $\approx 10^{12}$–10^{13} intersections for the grid tagging.

If the object is rigid and its position does not change in time with respect to the mesh, the tagging must be performed only once at the beginning of the simulation and a computational overhead of a few minutes does not impact the total cost of the whole simulation. In contrast, when objects are deformable or they move with respect to the grid, the tagging must be performed at every time step (or more than once in case iterations are needed, as for the fluid–structure interaction) and the computational overhead entailed by the computation of $N \times M$ intersections becomes unbearable.

An efficient modification of the aforementioned procedure consists of enclosing the real object of interest in a bounding-box whose faces, being six plane rectangles, can be described by a total of 12 triangles.

This bounding-box can then be used for a preliminary sifting of the mesh nodes to decide whether or not to perform the control against each surface triangle; in fact, if a grid point is inside the bounding-box it might also be inside the object, while those points outside the bounding-box are definitely also outside the object.

Since all the N mesh points must be tested against the 12 triangles of the bounding-box, $12N$ intersections are always necessary for the preliminary test. However, assuming that of all N grid points of the computational domain, only a fraction $p < 1$ is inside the bounding-box, then the additional intersections to be checked reduce to $pN \times M$, thus yielding a total number of $(12 + pM) \times N$. Comparing this value with $M \times N$ of the naive approach, it turns out that the bounding-box test is convenient whenever $p < (M - 12)/M \approx 1$, which is true in all practical cases. Furthermore, in case the bounding-box is aligned with the coordinate lines of a Cartesian mesh, the check can be performed in a straightforward way without resorting to the ray tracing method.

Obviously, the saving of computational time is the larger the smaller is p or, in other words, the smaller is the bounding-box with respect to the computational domain; this suggests that further gains can be obtained by splitting a complex object into a set of R partial elements, each one described by M_i triangles (with $M = \sum_{i=1}^{R} M_i$). In this case, we have a fraction p_i of the N nodes passing the bounding-box test for the ith body entailing a total number of intersections of $12RN + N \sum_{i=1}^{R} p_i M_i$: Savings with respect to the naive method are obtained when $\sum_{i=1}^{R} p_i M_i < M - 12R$ which, once again, is true for all realistic computations.

In order to make the previous discussion more quantitative, we consider the same racing car as in Figure 3.2, but now described by $M = 7.23 \times 10^5$ triangles, and a computational domain with $N = 2 \times 10^7$ nodes. The naive method for the grid tagging requires $M \times N = 1.45 \times 10^{13}$ intersections which,

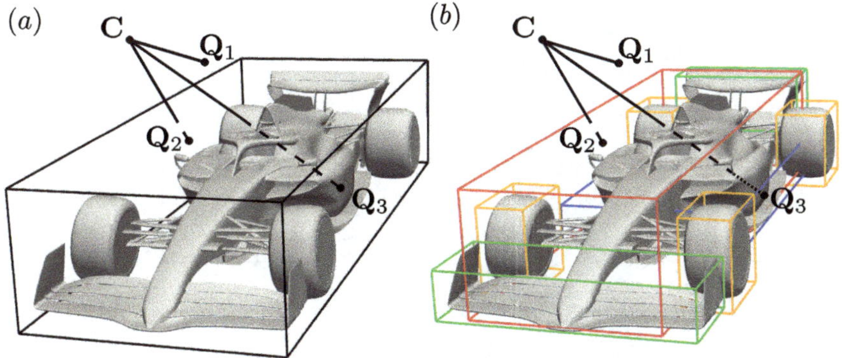

Figure 3.7 Bounding-box surrounding a complex object. The bounding-box test excludes the possibility for point Q_1 to be inside of the object without computing the intersections between the ray CQ_1 with all the triangles of the object.

on a single-core Apple M1 at 3.2 GHz, can be computed in 572 s. This time is incomparably shorter than that needed to build a body-fitted mesh and negligible when compared to that of the flow computation; therefore it would not be much of a problem for simulations dealing with steady rigid objects. The same computational load, however, would become unbearable should the tagging be necessary at every time step, as in the case of moving or deformable surfaces.

Enclosing the car in a bounding-box (Figure 3.7a), for the given domain size and grid distribution (with bounding-box and domain volumes, respectively, of $V_c \simeq 24.64$ and $V = 3000$, in arbitrary units) we have obtained $p \simeq 8.2 \times 10^{-3}$ resulting in a number of intersections of $(12 + pM)N = 1.2 \times 10^{11}$, which could be computed within 4.73 s on the same computational system as mentioned previously.

Finally, splitting the car into $R = 11$ partial objects with triangles ranging from $(M_i)_{\min} = 1.5 \times 10^4$ to $(M_i)_{\max} = 1.65 \times 10^5$, more refined bounding-boxes can be obtained (Figure 3.7b) yielding $p = \sum_{i=1}^{R} p_i = 5.8 \times 10^{-3}$ and a number of intersections $12RN + N \sum_{i=1}^{R} p_i M_i \simeq 1.1 \times 10^{10}$ which required only 0.43 s. This computational overhead is already small, compared to the integration of the Navier–Stokes equations, and it could be sustained at every time step without further changes. For simulations on parallel computers, multiple processors can be used to tag simultaneously different mesh blocks with respect to each partial elements, thus obtaining additional gains.

3.3.1 Open Surfaces

One of the pillar hypotheses at the base of the ray-tracing theory is that the surface must be "watertight" so that it encloses a finite volume. This is necessary

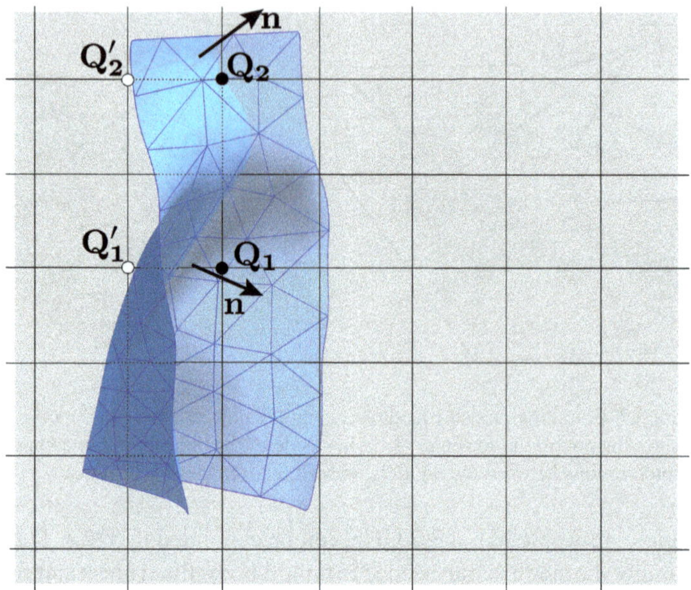

Figure 3.8 Determination of the interface points for an open surface: $\mathbf{n}$ is the local positive normal to the surface, $\mathbf{Q}_i$ and $\mathbf{Q}'_i$, $i = 1, 2$ are the "front" and "rear" external points, respectively.

in order for the domain to be split into inner and outer parts and to identify a fluid side. In many instances, however, the geometry has a negligible thickness when compared to the other in-plane dimensions and the structure can be described only by its surface. Of course, in reality, *zero-thickness* objects do not exist, although shells and membranes can be considered a good approximation and they have a large relevance in Nature and technology. Since these geometries do not have inner points, the ray-tracing theory cannot be applied to tag the grid points of the mesh. Nevertheless, the surface can always be described by triangles and the coordinates of the intersection with a ray can be found by using the same tools as for more standard geometries enclosing a finite volume.

In fact, in Chapter 4, it will be shown that the knowledge of the intersection and the closest external points can be sufficient to impose the desired boundary conditions on the immersed surface.

It is worth mentioning that even if for these structures it is not possible to define inner and outer points, it is still easy to distinguish to which side of the surface each external point belongs. This can be easily achieved considering that all surface triangles have a normal whose orientation depends on the sequence by which the vertices are assigned. If the surface triangulation is correctly performed, all the normals are consistent and the external points can be correctly oriented.

This information can be used to deal with surfaces "wet" by the fluid on both sides, like a flag fluttering in the wind (Huang and Sung, 2010), or for boundaries interacting with the flow on only one side, like a cell membrane transported by a liquid stream (Lu et al., 2024).

4

Forcing Methods

Once the geometry has been immersed into the grid and its computational nodes properly tagged, we must move forward with the definition of a procedure to locally modify the governing equations to account for the presence of the body. This is achieved by prescribing a distribution of fictitious forcing terms, $\mathbf{f}$, added in the Navier–Stokes equations:

$$\nabla \cdot \mathbf{u} = 0,$$

$$\frac{\partial \mathbf{u}}{\partial t} + \nabla \cdot (\mathbf{uu}) = -\frac{\nabla p}{\rho} + \nu \nabla^2 \mathbf{u} + \mathbf{f}. \tag{4.1}$$

The momentum equation already includes $\mathbf{f}$ to account for volume forces such as buoyancy and centrifugal force; in this context, there is an additional contribution to enforce the proper flow boundary condition at the immersed interface. The advantage of this approach is that $\mathbf{f}$ can be assigned on a structured regular grid and the need for a body-conformal mesh is avoided.

Different methods to calculate the forcing field $\mathbf{f}$ have been proposed in the literature and the main ones are illustrated in this chapter.

The flow charts of Figure 4.1 outline the rationale behind two alternative broad families of IBMs, namely *continuous* and *discrete*: In the former, the forcing is incorporated into the governing equations before discretization, whereas in the latter it is introduced only after the equations have been discretized on a grid.

Within the first approach, $\mathbf{f}$ is defined at the body surface ("sharp continuous IB forcing" in Figure 4.1a) to be diffused around the fluid–solid interface ("smooth continuous IB forcing"), which is then included as a volume forcing in the Navier–Stokes equations. The resulting continuous system of equations is discretized over a structured Cartesian mesh and solved numerically.

On the other hand, when a discrete IBM is employed, the Navier–Stokes equations are first discretized in a domain containing the immersed bodies, neglecting their presence. The discretization scheme at the interface cells is successively adjusted ("boundary reconstruction method") to account for the action of the body on the fluid. In particular, since the flow is generally solved on a fixed regular grid, not aligned with the body surface, the no-slip boundary conditions are then imposed by appropriate interpolation rules preserving the order of accuracy of the numerical scheme.

An advantage of continuous IBMs is that the scheme is largely independent of the underlying spatial discretization and such an approach can be quickly implemented into an existing Navier–Stokes solver by simply adding a forcing term. However, continuous approaches rely on smoothing functions that smear the solid–fluid interface over several cells across the boundary; this entails additional grid refinement for the proper location of the boundary, which exacerbates the resolution problems of CFD simulations with IBMs.

On the other hand, discrete IBMs change the discretization scheme of the equations only at the interface cells, resulting in a sharper representation of the immersed boundary. This is a desired feature, especially in high Reynolds number flows, since thin boundary layers already stress the resolution capability of non-body-conformal meshes and additional burdens related to the numerical method are hardly bearable. The discrete approach, while dependent on the discretization method, can be easily integrated into existing finite-difference or finite-volume structured solvers, preserving the code's programming ease and computational efficiency.

4.1 Continuous Forcing

4.1.1 IBM with Elastic Boundaries

One of the first implementations of IBMs dates back to the early 1970s, when Peskin (1972a) solved numerically the interaction between the blood flow through a mitral valve and its deformable leaflets in an idealized two-dimensional setting. In that formulation, solid boundaries were modeled by a collection of massless elastic fibers, described by their curvilinear path $\mathbf{X}(s, t)$ at each time t with s the arc length; these fibers were immersed in a uniform Cartesian grid, where the incompressible Navier–Stokes equations could be discretized using a finite-difference method.

In the following, we will use capital letters to refer to Lagrangian quantities and the same symbols in lower case for the Eulerian counterparts. A number

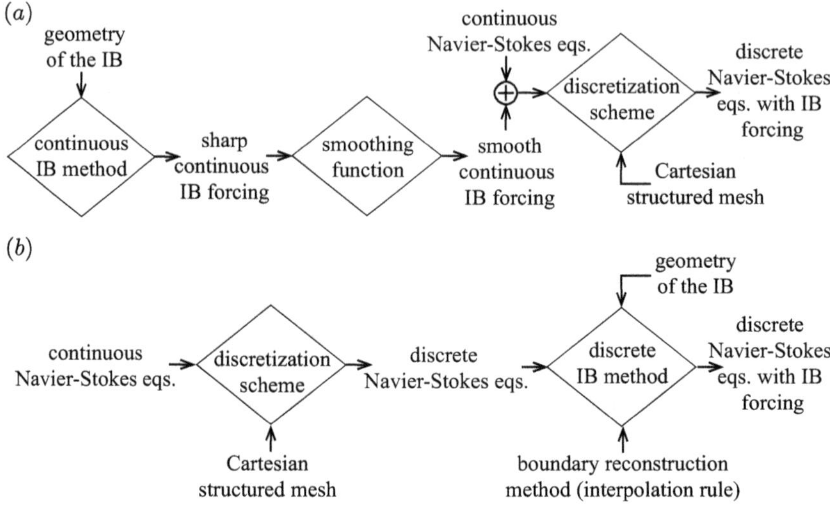

Figure 4.1 Schematic of the (a) continuous and (b) discrete IBM.

of Lagrangian markers $\mathbf{X}_i(t) = \mathbf{X}(s_i, t)$ is attached to the fibers and they move with the local fluid velocity $\mathbf{u}$ following:

$$\frac{d\mathbf{X}_i(t)}{dt} = \mathbf{u}[\mathbf{X}_i(t)]. \tag{4.2}$$

Note that fluid velocity is defined on the Eulerian mesh and its value at the Lagrangian marker $\mathbf{u}[\mathbf{X}_i(t)]$ can only be obtained via interpolation from the surrounding nodes.

The relative displacement of two neighboring Lagrangian markers $\mathbf{X}_i$ and $\mathbf{X}_j$ produces internal stress in the fiber, which depends on the material constitutive law

$$\mathbf{F}(s, t) = -\frac{\partial E}{\partial \mathbf{X}}, \tag{4.3}$$

with E the functional of the elastic potential energy of the instantaneous configuration $\mathbf{X}(s, t)$; for the linear elastic materials of Peskin (1972a), equation (4.3) reduces to Hooke's law.

In the case of a massless boundary, hydrodynamic loads at each material point are in balance with the internal stress of the fiber; thus, in principle, the

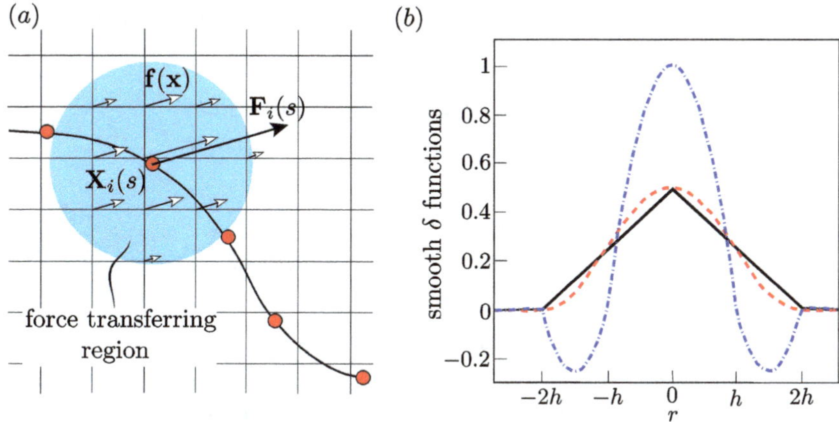

Figure 4.2 (a) Transfer of force density $\mathbf{F_i}$ from a Lagrangian element $\mathbf{X}_i(s)$ to the surrounding Eulerian grid. The shaded region indicates the transfer region given by the mollified δ-function. (b) Mollified δ-functions: red dashed line – – – – for equation (4.5) proposed by Peskin (1972a), black solid line ——— for (4.6) proposed by Saiki and Biringen (1996) and blue dot-dash line —·— of (4.7) introduced by Beyer and LeVeque (1992).

forcing in the momentum equation (4.1) exerted by the fibers on the fluid could be expressed through

$$\mathbf{f}(\mathbf{x}, t) = \int_{\partial B} \mathbf{F}(s, t)\delta[\mathbf{x} - \mathbf{X}(s, t)]\mathrm{d}s, \tag{4.4}$$

where ∂B is the immersed boundary and δ the Dirac delta-function.

Unfortunately, an expression like (4.4) can not be used to obtain $\mathbf{f}$ on the computational mesh as the fibers generally do not cross the Eulerian nodes of the Cartesian grid. The problem is of the same nature as for equation (4.2), although in that case $\mathbf{u}$ was available on the grid and needed on the markers; here $\mathbf{F}$ is known on the markers and needed on the grid. Moreover, the introduction of a Dirac δ-function at the immersed interface produces a strong discontinuity, resulting in numerical oscillations (wiggles) and a degraded spatial accuracy of the method (Peskin, 1972a).

In order to avoid these issues, the real sharp δ-function is replaced by a smooth surrogate spreading the forcing over a transition region. This is equivalent to spreading the δ-function itself over a narrow band (typically three or four nodes) across the boundary, so that each Lagrangian marker will have a force transferring region with the Eulerian nodes therein having a share of the momentum force density (see Figure 4.2a). The choice of the mollified δ for the distribution $\mathbf{f}$ is a key ingredient in this method and several alternatives have been proposed:

$$\delta(r) = \frac{1}{h} \begin{cases} (\cos(\pi r/2h) + 1)/4h, & \text{if } r \leq 2h, \\ 0 & \text{otherwise,} \end{cases} \tag{4.5}$$

or

$$\delta(r) = \frac{1}{h} \begin{cases} (2h - r)/4h^2, & \text{if } r \leq 2h, \\ 0 & \text{otherwise,} \end{cases} \tag{4.6}$$

or

$$\delta(r) = \frac{1}{h} \begin{cases} 1 - (r/h)^2, & \text{if } r \leq h, \\ 2 - 3r/h + (r/h)^2, & h \leq r \leq 2h, \\ 0 & \text{otherwise,} \end{cases} \tag{4.7}$$

where h is the grid spacing and r is the distance between the Lagrangian marker of the fiber and any Eulerian point. Figure 4.2b shows some smooth δ-functions: Expression (4.5) is the original transfer function introduced by Peskin (1972a), while that of equation (4.6) was suggested by Saiki and Biringen (1996). Both expressions were shown to yield a first-order accurate solution on account of the presence of a force discontinuity. Equation (4.7) was introduced by Beyer and LeVeque (1992) and it was shown to yield a second-order accurate solution even in the presence of discontinuities.

It should be noted that regardless of the particular expression for $\delta(r)$, all the above procedures smear the immersed boundary over the nodes surrounding the body–fluid interface. At low Reynolds numbers, the extent of the smearing can be small when compared to the thickness of the viscous boundary layer and the effect on the flow dynamics is negligible. On the other hand, even if this smearing can be reduced by clustering the nodes in the boundary region, the uncertain boundary location becomes a serious limitation of the method, even at moderately high Reynolds numbers.

A final word of warning is about the applicability of this method to solid undeformable bodies: The flexibility of the boundaries is modulated by the functional E of (4.3) or, for linearly elastic materials, just by the elastic coefficient κ. In principle, rigid boundaries could be obtained in the limiting case of infinite stiffness ($\kappa \rightarrow \infty$), which clearly can not be used in a numerical algorithm. As an approximation of the above condition, very high values of κ are employed even if it is observed that reducing the deformability of the boundaries stiffens the governing equations, which need smaller time steps for numerical integration. The simulation of solid bodies is therefore a compromise between rigid enough boundaries and small enough stiffness constants to make the problem tractable.

4.1.2 The Feedback Forcing

A very common problem is that of a flow interacting with an object for which the velocity at the interface $\mathbf{V}(s,t)$ is either known as a boundary condition, if the body is at rest or moving with imposed kinematics, or is obtained from the solution of Newton–Euler equations (see Chapter 7). In this case, the force density of equations (4.1) can be designed directly to enforce the condition $\mathbf{V} \equiv \mathbf{u}$ at the immersed boundary. Another difference with respect to Section 4.1.1 is that now $\mathbf{V}$ is available for any values of the arc length s and therefore there is no need to distribute Lagrangian markers on the boundary and track their position in time.

Goldstein et al. (1993) and Saiki and Biringen (1996) proposed the following expression for $\mathbf{f}(\mathbf{x}_s,t)$:

$$\mathbf{f}(\mathbf{x}_s,t) = \alpha_f \int_0^t [\mathbf{u}(\mathbf{x}_s,t') - \mathbf{V}(\mathbf{x}_s,t')]\mathrm{d}t' + \beta_f [\mathbf{u}(\mathbf{x}_s,t) - \mathbf{V}(\mathbf{x}_s,t)], \quad (4.8)$$

with $\mathbf{x}_s = \mathbf{X}(s,t)$ and α_f, β_f *negative* constants.

We are confronted again with the problem that Eulerian quantities are available only at the mesh nodes whose positions do not coincide with the immersed boundary: $\mathbf{u}(\mathbf{x}_s)$ will therefore be obtained by interpolations from the surrounding nodes while $\mathbf{f}(\mathbf{x}_s)$ will be transferred back to the nodes interpolating from values available on the boundary.

The above expression can be interpreted as a feedback to the velocity mismatch at the interface $\mathbf{u}(\mathbf{x}_s) - \mathbf{V}(\mathbf{x}_s)$ and the negative constants are such that to enforce $\mathbf{u} \to \mathbf{V}$ on the immersed boundary. The second term can be interpreted as the resistance opposed by the boundary element to assume a velocity $\mathbf{u}$ different from $\mathbf{V}$.

An intuitive argument for understanding the action of the above forcing is obtained by retaining from equation (4.1) only the time derivative term on the left-hand side and the forcing on the right-hand side to obtain

$$\frac{\mathrm{d}\mathbf{q}}{\mathrm{d}t} \approx \mathbf{f} = \alpha_f \int_0^t \mathbf{q}\mathrm{d}t' + \beta_f \mathbf{q}, \qquad (4.9)$$

with $\mathbf{q} = \mathbf{u} - \mathbf{V}$. Equation (4.9) represents a simple damped oscillator with frequency $(1/2\pi)\sqrt{|\alpha_f|}$ and damping coefficient $-\beta_f/(2\sqrt{|\alpha_f|})$. This implies that as $\mathbf{u}$ on the boundary becomes different from $\mathbf{V}$, the forcing $\mathbf{f}$ restores $\mathbf{u}$ back to $\mathbf{V}$. In an unsteady flow, the magnitude of α_f must be large enough so that the restoring force can react with a time scale shorter than any flow variation. Unfortunately, the value of these constants is flow-dependent and, even if for sufficiently large α_f and β_f the flow becomes independent of their value, there is no general rule for their determination, so ad hoc judgments are needed. Furthermore, the major drawback of this forcing is that large values of α_f and β_f make the equations stiff, and the time integration requires very small time steps. According to the stability analysis of Goldstein et al. (1993), when all the IB forcing terms are computed explicitly with an Adams–Bashforth scheme, the stability limit is given by $\Delta t < \left(-\beta_f - \sqrt{\beta_f^2 - 2\alpha_f \kappa}\right)/\alpha_f$ (with $\kappa \sim O(1)$ a flow-dependent constant), which is too restrictive for three-dimensional flows in complex geometries (Goldstein et al., 1993).

A partial overcoming of this stability issue is to treat the second term on the right-hand side of equation (4.9) implicitly in time. As an example, if an Adams–Bashforth scheme is used for the explicit term and a Crank–Nicolson scheme is used for the implicit term the time-discrete forcing reads

$$\mathbf{f}^{n+1/2} = \alpha_f \left[\frac{3}{2} \int_0^{t^n} \mathbf{q} dt' - \frac{1}{2} \int_0^{t^{n-1}} \mathbf{q} dt' \right] + \beta_f \left[\frac{\mathbf{q}^{n+1} + \mathbf{q}^n}{2} \right], \qquad (4.10)$$

where n is the discrete time level, with the forcing evaluated at the intermediate time $n + 1/2$ to retain the second order accuracy of the time derivative.

Fadlun et al. (2000) have shown that this modification indeed improves the time step limitation, although its size still is strongly dependent on the coefficients α_f and β_f as well as on the specific flow and its Reynolds number; see Figure 4.8, along with the discussion that follows.

4.1.3 Penalty Methods

This class of methods (often referred to as fictitious-domain or domain-embedding methods) models the volume of the solid body as a porous medium and solves the corresponding Navier–Stokes equations (4.1) with the forcing $\mathbf{f}$ consisting of a Darcy drag term. In more detail, the following *penalty* function nudging the Eulerian velocity $\mathbf{u}$ in the body towards the target value $\mathbf{V}$ is added in the momentum equation:

$$\mathbf{f}(\mathbf{x}, t) = \frac{\nu[\mathbf{u}(\mathbf{x}, t)) - \mathbf{V}(\mathbf{x}, t))]}{\rho K}, \qquad (4.11)$$

where K is a free parameter, modulating the permeability of the porous medium immersed in the flow (or to the Darcy number in nondimensional form). If $K \longrightarrow \infty$ the forcing vanishes and the standard unforced Navier–Stokes equations are retrieved. In contrast, if $K \longrightarrow 0$ the forcing becomes dominant, the porous medium behaves as a solid body and the flow around an arbitrary shaped object can be obtained with the velocity $\mathbf{u} = \mathbf{V}$ within the body.

In equation (4.11) the target velocity $\mathbf{V}$ is defined directly at the same grid point $\mathbf{x}$ as the Eulerian counterpart. This is not a problem for the nodes inside the body as the velocity field is known at any location, for example as for rigid-body motion. On the other hand, at the interface, there is the usual problem of the boundary not crossing the mesh nodes; therefore, $\mathbf{V}$ must be obtained by interpolation.

This method looks similar to the feedback forcing of Section 4.1.2 (they are essentially equivalent when choosing $\alpha_f = 0$ and $\beta_f = \mu/K$), although it has a different physical interpretation, opening the possibility to a new type of application. In fact, for $0 < K < \infty$ the forcing can be modulated to provide a momentum loss in a desired region, thus simulating porous media. Furthermore, the vector form of (4.11) allows us to use different penalization constants K in different directions to obtain anisotropic momentum losses, as in the case of a porous medium having channels with a preferential orientation.

Equations (4.1) with the forcing (4.11) become the Navier–Stokes–Brinkman equations that can be solved over the whole domain with different values of K depending on the zonal characterization (fluid, solid or porous medium). This approach allows us to simulate the presence of a grid or other elements producing momentum and energy losses without simulating explicitly all the details of the objects, which in some circumstances would be exceedingly difficult or impossible (like for the flow behind all the branches and leaves of a tree).

A drawback of the method is that in numerical simulations the value of K can be neither 0 nor ∞ and solid or fluid regions are approximated by finite, user-specified values, which should be tuned for each case depending on the specific flow features and Reynolds number. The main issue arises from using very small values of K, which increase the stiffness of the governing equations and consequently degrade the convergence properties of the solution procedure. The final values of K must be, once again, a compromise between the need to approximate solid boundaries and the preservation of numerical stability at a reasonable computational cost.

4.2 Discrete Forcing

In discrete methods, the IB forcing acts directly on the discretized Navier–Stokes equations by locally adjusting the numerical scheme in the proximity of the body surface. A further distinction within this class of IBMs is made depending on whether the IB forcing is computed directly at the Eulerian points close to the immersed body (*Eulerian* methods) or at the control points lying on the Lagrangian mesh and then transferred to the Eulerian mesh (*Lagrangian methods*).

4.2.1 Eulerian Methods

We consider the momentum equation (4.1) with the time derivative in discrete form:

$$\frac{\mathbf{u}^{n+1} - \mathbf{u}^n}{\Delta t} = \mathbf{rhs}^{n+1/2} + \mathbf{f}^{n+1/2}, \tag{4.12}$$

where $\mathbf{rhs}^{n+1/2}$ contains convective and viscous terms along with the pressure gradient.

If now we ask which value of $\mathbf{f}^{n+1/2}$ will yield $\mathbf{u}^{n+1} = \mathbf{V}^{n+1}$ on the immersed boundary, the answer is simply given by

$$\mathbf{f}^{n+1/2} = -\mathbf{rhs}^{n+1/2} + \frac{\mathbf{V}^{n+1} - \mathbf{u}^n}{\Delta t} \quad \text{on the boundary, } \partial B. \tag{4.13}$$

As originally proposed by Mohd-Yusof (1997), this forcing is *direct* in the sense that the target velocity value is imposed directly on the boundary without the mediation of any dynamical processes and the boundary condition is satisfied at every time step regardless of the time scales of the flow.

This procedure does not require the introduction of singular functions or mollified surrogates, nor does it need user-defined parameters (as for the continuous forcings), thus the boundary condition is sharply imposed at the fluid/body interface without spreading the IB forcing.

Finally, equation (4.13), plugged into (4.12), yields directly $\mathbf{u}^{n+1} = \mathbf{V}^{n+1}$ which can be directly imposed along the immersed boundary without explicitly computing $\mathbf{f}$, unless needed for other reasons.

Boundary Reconstruction

The forcing equation (4.13) assumes the coincidence of the Eulerian node, where $\mathbf{u}$ is defined, and the point of the immersed boundary where $\mathbf{V}$ is known; unfortunately, this is seldom the case and also for this method some interpolation is needed. Furthermore, even if the immersed boundary were completely coincident with coordinate lines, the staggered discretization of the variables, widely adopted for incompressible Navier–Stokes equations, would unavoidably imply interpolations for those velocity components not lying exactly on the boundary.

In fact, since the mesh is not body conformal, the final goal of any IB method is to enforce a velocity $\mathbf{V}$ over a boundary, with the flow unknowns (velocity

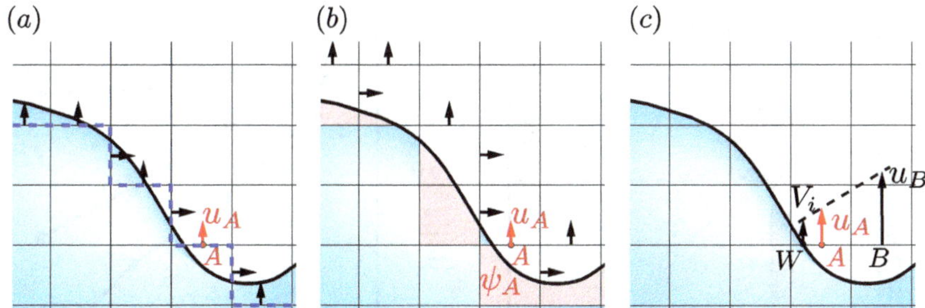

Figure 4.3 Sketch of the interpolation procedures: (a) no interpolation corresponding to a stepwise treatment of the geometry, (b) volume fraction weighting, (c) velocity interpolation. In (a) A is the point closest to the body surface, whereas in (b) and (c) it is the closest external point. In (a) and (b) $u_A = V_i$, while in c) B and W indicate the second external grid point and the location of the body surface crossed by the computational grid, respectively.

components, pressure or scalars) located at close but different positions. This entails some modifications of the algorithm at the immersed surface referred to as "boundary reconstruction," whose detail can determine the overall accuracy of the numerical scheme.

Following Fadlun et al. (2000), the simplest possibility is to select the grid nodes closest to the immersed boundary (such as point A in Figure 4.3a) and apply the forcing therein, as if that point were on the boundary. In this case, for the vertical velocity component we write $u_A = V_i$ and the corresponding component (say the ith) of the IB forcing reads

$$f_i^{n+1/2} = -\text{rhs}_i^{n+1/2} + \frac{V_i^{n+1} - u_i^n}{\Delta t} \quad \text{on the closest node, } A. \quad (4.14)$$

Obviously, in this case, there is no interpolation and the real boundary is described by a stairstep approximation (see Figure 4.3a) whose jumps reduce as the Eulerian mesh gets more refined.

A second possibility consists of computing for each cell crossed by the boundary (interface cell) the volume fraction occupied by the body and the total volume of the cell: Their ratio ψ is then used as a weight coefficient to scale the forcing applied to the external grid points closest to the boundary. For example, with reference to Figure 4.3b, the momentum equation for the vertical velocity component at point A becomes

$$f_i^{n+1/2} = \psi_A \left(-\text{rhs}_i^{n+1/2} + \frac{V_i^{n+1} - u_i^n}{\Delta t} \right) \quad \text{on the closest external node, } A.$$

$$(4.15)$$

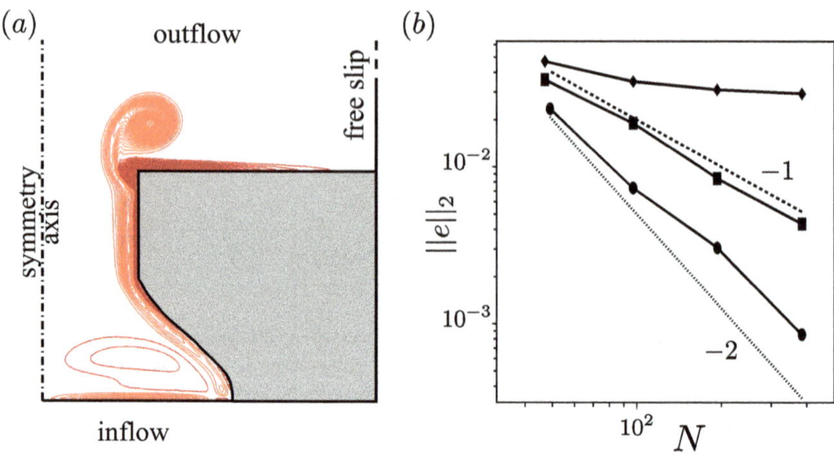

Figure 4.4 (a) Azimuthal vorticity contour plots for the vortex ring formation from a curvilinear nozzle at $Re = 1500$ (129×257 grid points in the radial and axial directions). (b) L_2-norm error of the axial velocity component versus number of grid points: • interpolated velocities, ■ volume fraction, ♦ stepwise geometry; ········· -2 slope, $----$ -1 slope.

In the third case, the same velocity component in A is computed considering that point in between W at the boundary and B completely within the fluid: The velocity in the latter is given by the governing equations, while in the former V_i is known as the boundary condition. It is therefore possible to compute the velocity in A as

$$u_A = \phi_B u_B + \phi_W V_i, \tag{4.16}$$

where $\phi_{B,W}$ are interpolation coefficients in a linear approximation (Figure 4.3c).

The boundary condition is thus enforced not directly at the wall but by imposing the interpolated velocity value (4.16) in A using the following forcing:

$$f_i^{n+1/2} = -\text{rhs}_i^{n+1/2} + \frac{u_A^{n+1} - u_i^n}{\Delta t} \quad \text{on the closest external node, } A. \tag{4.17}$$

It is worth mentioning that when the velocity is discretized on a staggered mesh, as in Figure 4.3, each component has different interface cells, volume fractions occupied by the body and interpolation coefficients for the imposition of boundary conditions.

Fadlun et al. (2000) compared the three reconstruction procedures using the transient flow through a curvilinear nozzle during the formation of the starting vortex ring (Figure 4.4a). Forcings (4.14)–(4.17) were combined with the

Navier–Stokes solver in cylindrical coordinates of Verzicco and Orlandi (1996) based on a second-order finite-difference discretizations on a staggered mesh. Axisymmetric simulations were run on grids of increasing resolution and that on the finest was regarded as the exact reference solution. The corresponding errors in L_2 norm of the axial velocity components are reported in Figure 4.4b, showing different behaviors for the three reconstructions illustrated previously.

The stairstep geometry yields an error decreasing more slowly than first order since, even if the geometry smoothens, the wall velocity gradients become more jerky as the grid is refined and eventually the error saturates to a constant value. Better results are obtained by weighting the forcing by the volume fraction of the cell occupied by the body, improving the error to first order. In this case, one problem is that regardless of how fine the mesh is, the forcing weight ψ jumps along the interface cells (see Figure 4.4b) and the discontinuous distribution of the forcing degrades the accuracy of the solution. Furthermore, simulations of flows over immersed flat surfaces have shown that this interpolation procedure underestimates the velocities at the boundary; therefore, even if the geometry is not stepwise the velocity boundary conditions are not completely satisfactory yet.

The best results, both in terms of error magnitude and accuracy, are obtained by the third method, which yields essentially a second-order accurate solution. This is because of the linear interpolation of the velocity u_A, which is second-order itself and provides a boundary reconstruction consistent with the underlying numerical scheme sharing the same accuracy.

The second-order accuracy of the linear reconstruction is a desirable feature, even if it must be noted that with this procedure a linear velocity profile is imposed between the wall and the first external grid point, rather than solving the governing equations. Such an approach requires the grid to be fine enough in the wall region for the first node to be in the viscous sublayer[1] and this is not generally possible for every complex geometry and for high Reynolds number flows, as discussed in Chapter 8.

Despite the second-order accuracy, the linear reconstruction of Figure 4.4c is still quite primitive and it can be improved on several counts. For example, an objective criterion could be set to decide whether the interpolation at point A should be performed along the vertical direction rather than the horizontal one (or both). Another possibility is to consider also the first internal point for the interpolation or choose the one nearest to the wall between the first internal and the first external. Also, the wall-normal information could be included in

---

[1]The viscous sublayer is the part of the boundary layer closest to the wall where tangential velocity increases linearly with the wall distance. Its thickness depends on the laminar or turbulent nature of the flow, although it is a smaller fraction of the boundary layer thickness, the higher is the Reynolds number (Schlichting, 1968).

the reconstruction process and, finally, polynomials of higher order than linear could be used for more flexible interpolations.

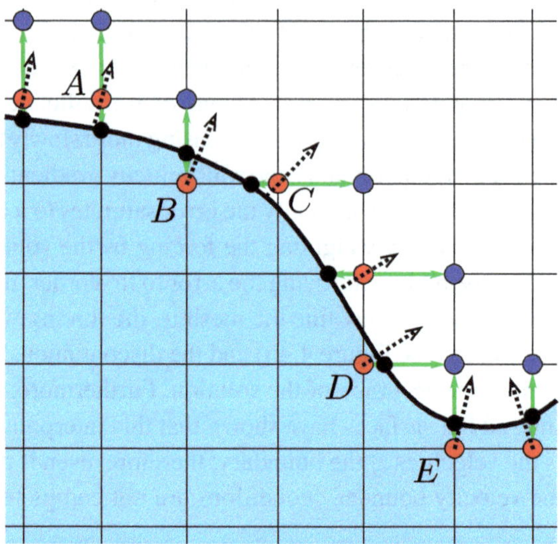

Figure 4.5 Boundary reconstruction scheme imposing IB forcing at the node nearest to the boundary as in Giannetti and Luchini (2007). The – – – – arrow is the local outward boundary normal crossing the closest node; red bullets are forced points; blue bullets and black bullets indicate, respectively, fluid and boundary points used for interpolation/extrapolation. The thick green double arrow is the stencil used for reconstruction.

According to the above observations, several improvements have been proposed in the literature and here we describe a few of them with the aim of indicating the path to build more sophisticated reconstructions.

Giannetti and Luchini (2007) suggest that, rather than imposing the IB forcing always at the first external point, it can be applied at the point closest to the immersed surface, which can be either internal or external. The neighboring external fluid point is then used for the interpolation, in the vertical or horizontal direction depending on which is closer to the local normal. An example of the corresponding nodes involved in the boundary reconstruction is shown in Figure 4.5; it can be observed that, depending on the crossing of the boundary with the coordinate lines, the forced node switches from external to internal and the velocity therein is obtained, respectively, by interpolation or extrapolation with the boundary and the fluid values.

A possible weakness of this procedure is that the interpolation direction jumps discontinuously depending on the local orientation of the boundary (see points *B* and *C* of Figure 4.5). This is because of the particular reconstruction

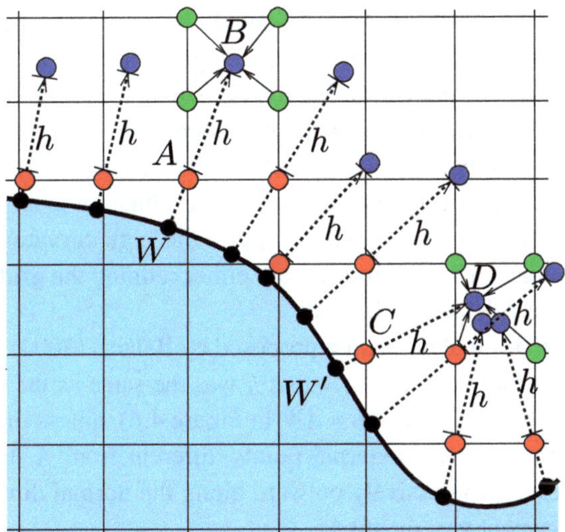

Figure 4.6 Boundary reconstruction scheme along the wall-normal direction as proposed by Balaras (2004). The $----$ arrow is the local outward normal direction crossing the closest node; red bullets are forced points; black bullets indicate boundary points; blue bullets are probe points used for the interpolation and green bullets are fluid points from which probe values are computed.

method using information only from a single direction of the coordinate lines, which can be acceptable when the immersed boundary is largely aligned with the grid but certainly not for a general geometry.

Indeed, for highly curved boundaries, the choice of the reconstruction direction may not be unique and the same fluid point can be used more than once for interpolation (like points D and E of Figure 4.5). Such interpolation ambiguities were addressed by Balaras (2004) through a boundary reconstruction method based on the local outward normal, which is well-defined regardless of the relative orientation of boundary and grid lines. Following this idea, the closest external grid points are identified (red bullets in Figure 4.6) along with the corresponding boundary points (black bullets) given by the intersection of the normal direction (dashed lines) passing through the external node and the boundary. Then, for each external point, moving along the wall-normal for a fixed distance h, a fluid "probe" is selected that usually does not coincide with any grid point. Using the surrounding fluid nodes, however, the fluid velocity on the probe can be computed and then used for linear interpolation with the boundary condition to reconstruct the value at the external point.

For this method, the choice of the probe distance h is quite important since it should be small enough to keep the wall reconstruction local but not too small for probe values to be "contaminated" by interpolated points; usually values $1.5\Delta \leq h \leq 2.5\Delta$ are used, with Δ being the local grid size. The optimal configuration is that of point A of Figure 4.6, where the probe B is surrounded only by fluid nodes and values in A can be obtained by interpolating between B and W. In contrast, point C is a critical one since values in its probe D depend on a reconstructed point; the occurrence of this situation is an indication that the underlying mesh is too coarse to capture the high curvature detail of the immersed boundary. The possibilities are either refining the grid or smoothing out the small-scale details of the body.

Indeed, in the original procedure proposed by Balaras (2004) the position of the probe was initially defined such that h was the same as the distance of the external point from the wall ($AB = AW$ in Figure 4.6), unless the interpolation for the probe involved any external points different from A itself. In such a case B was moved progressively outward along the normal direction until the previous requirement was satisfied.

Yet another approach is that proposed by Kang et al. (2009) who, for each external point A, built a triangle as in Figure 4.7a having as vertices B, C fluid nodes and W at the intersection of the wall-normal through A with the boundary. Assuming a linear distribution next to the wall for any velocity

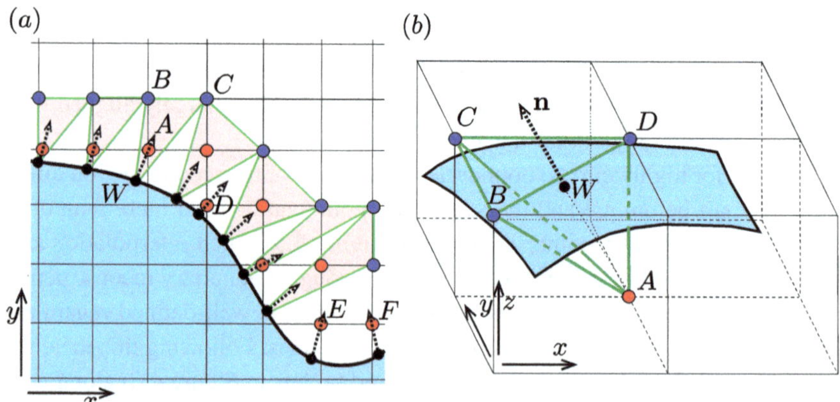

Figure 4.7 Sketch of possible interpolation schemes for geometries not aligned with the grid directions in (a) two and (b) three dimensions.

component and using a local coordinate system with the origin in W results in $u = p_1 x + p_2 y + u_W$ with the coefficients p_1 and p_2 determined imposing the known velocity values in B, C.

Once the coefficients are computed, the velocity component at the forcing point A reads

$$u_A = p_1 x_A + p_2 y_A + u_W, \quad \text{or} \quad u_A = \phi_B u_B + \phi_C u_C + \phi_W u_W,$$

with $\phi_{B,C,W}$ the interpolation coefficients, which can be used in the forcing equation (4.17).

Note that this method requires that for any forced point, a triangle can be found with two fluid vertices and a boundary point. Depending on the local grid size and boundary curvature, however, there can be points for which this is not possible, like points D, E and F of Figure 4.7a, where the fluid vertices of the corresponding triangle coincide with some neighboring external points.

Possible alternatives are a one-dimensional reconstruction along the normal or using a triangle with one of the vertices that is a reconstructed point; as long as these points are a small fraction of the total forced points this is not a problem, otherwise the underlying grid has to be refined for a proper description of the geomctry.

Finally, this method has been extended to three-dimensional geometries and slightly modified by Cristallo and Verzicco (2006) who considered the setup of Figure 4.7b with the boundary condition to be satisfied in W and the forced node A *inside* the body. In more detail, the outward normal to the immersed surface is cast from point A and the intersection W is determined. A tetrahedron containing W is then constructed using the node A and the three closest external nodes (B, C and D) on the other side of the boundary. It is worthwhile to point out that in this case all external points are fluid and the exceptions of Figure 4.7a do not occur. Furthermore, velocities of all external points are determined by Navier–Stokes equations since the reconstructed ones are inside the body where the solution is usually disregarded.

Again, a local reference frame with the origin in W and a linear velocity distribution yields $u = p_1 x + p_2 y + p_3 z + u_W$, with the coefficients determined by imposing the known velocity values:

$$
\begin{aligned}
u_B &= p_1 x_B + p_2 y_B + p_3 z_B + u_W, \\
u_C &= p_1 x_C + p_2 y_C + p_3 z_C + u_W, \\
u_D &= p_1 x_D + p_2 y_D + p_3 z_D + u_W.
\end{aligned}
\tag{4.18}
$$

Once u_A is computed, it can be used for the forcing of equations (4.17), but applied to the closest internal node.

A similar approach can also be adopted to impose velocity boundary conditions other than no-slip (Dirichlet): For example, if a Neumann boundary condition needs to be imposed at the wall:

$$\left. \frac{\partial u}{\partial n} \right|_W = \alpha_W,$$

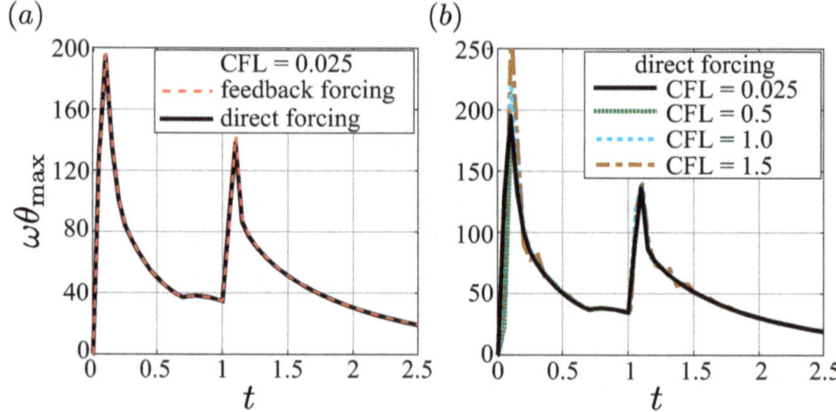

Figure 4.8 Time evolution of the peak azimuthal vorticity for the vortex ring formation problem of Fadlun et al. (2000). (a) Simulations at $CFL = 0.025$ for feedback (dashed line) and direct forcing (solid line). (b) Simulations with direct forcing for several CFL numbers (129×257 grid points in the radial and axial directions).

it is convenient to start from $u_A = p_1 x + p_2 y + p_3 z + p_4$, with four coefficients rather than three. Three conditions are given by (4.18) (with p_4 in place of u_W) while the fourth is

$$\left. \frac{\partial u}{\partial n} \right|_W = \nabla u \cdot \mathbf{n}|_W = p_1 n_x + p_2 n_y + p_3 n_z = \alpha_W, \qquad (4.19)$$

where n_x, n_y and n_z are the components of the outward normal vector in W. The resulting interpolated velocity component u_A is then directly imposed in A using the forcing (4.17).

Alternatively, rather than imposing a velocity value in a grid point, the no-slip condition at solid walls can be indirectly enforced by redefining the metrics for the flow derivatives at the first layer of interface fluid points, as done in Orlandi and Leonardi (2006) for the viscous terms.

Comparison of Direct Forcing with Feedback Forcing

Before concluding this section on Eulerian forcings, we briefly discuss the issue of efficiency, which greatly discriminates about the applicability of each method.

Here we report the results of Fadlun et al. (2000) who compared the direct forcing method (4.17) with the semi-implicit continuous approach described in Section 4.1.2: Both procedures were employed to simulate the vortex ring formation from a curvilinear nozzle shown in Figure 4.4a.

All simulations were performed using an adaptive time step Δt varied during the run to keep the stability parameter CFL constant (where CFL represents the Courant–Friedrichs–Lewy number). Starting from the continuous IB forcing (4.10), a series of simulations was executed with the CFL number gradually increased in order to determine the corresponding stability limit. It turned out that with the semi-implicit treatment of the forcing, simulations could be run up to $CFL = 0.025$, which is anyway a substantial improvement with respect to the stability limit of $CFL = O(10^{-3})$ observed in the case of fully explicit treatment.

The same simulation with $CFL = 0.025$ was then run using the direct forcing (4.13), obtaining a perfect agreement between the two as reported in Figure 4.8a: The evolutions of the peak azimuthal vorticity of the vortex are indistinguishable from each other. However, while the continuous forcing is already at its stability limit, the simulation with the discrete forcing can be run at much higher CFL without losing stability or accuracy. In Figure 4.8b, the same quantity as Figure 4.8a is shown for several values of CFL. It appears that the long-term dynamics are well predicted even using $CFL = 1.5$, which is very close to the theoretical stability limit ($CFL = \sqrt{3}$) of the third-order Runge–Kutta, used for the time advancement of the equations (Verzicco and Orlandi, 1996). The speed-up factor of about 60 of the direct forcing makes this technique very convenient for the computation of three-dimensional flows.

It must be pointed out that the above numbers are dependent on the specific problem and better computational performance, from the feedback forcing, could be obtained by dynamically tuning the constants α_f and β_f during the flow evolution. However, since a quantitative procedure supported by theory is not available, such an approach would rely on an empirical and user-dependent judgment, which is not desirable in a robust numerical procedure.

4.2.2 Lagrangian Methods

The direct forcing methods of Section 4.2.1 have been revealed as very efficient and attractive for the simulation of flows around fixed rigid bodies. As noted by Uhlmann (2005), however, interpolation procedures relating values at fixed grid nodes with values at arbitrarily located boundaries can lead to local force oscillations, which are undesirable, especially when the body dynamics is connected directly with this force.

In order to make up for this problem, Uhlmann (2005) has conceived a method in between the direct forcing of Fadlun et al. (2000) and that with Lagrangian markers of Peskin (1972a), taking the best of both. The main idea is to compute the forcing directly on the markers, evenly distributed on the immersed surface, using smooth transfer functions with the fixed Cartesian mesh. It is important to note that, differently from Peskin (1972a), in this context the markers (referred to as "force points") follow the motion of the body and they are not advected with the local fluid velocity. This implies that they are not additional degrees of freedom of the system.

The key ingredient of the method is a kernel to transfer quantities back and forth between Eulerian nodes and force points; since the performance of the method is quite dependent on the details of this transfer process, this will be discussed separately in the next two subsections.

To describe the algorithm step by step, we again write the time-discrete momentum equation,

$$\frac{\mathbf{u}^{n+1} - \mathbf{u}^n}{\Delta t} = \mathbf{rhs}^{n+1/2} + \mathbf{f}^{n+1/2}, \tag{4.20}$$

which, pretending the target boundary velocity $\mathbf{V}$ is located at the same point as the Eulerian node, could be solved for the forcing as

$$\mathbf{f}^{n+1/2} = -\mathbf{rhs}^{n+1/2} + \frac{\mathbf{V} - \mathbf{u}^n}{\Delta t}. \tag{4.21}$$

On the other hand, without inserting the forcing (4.21) in equation (4.20) the resulting velocity would be different (say $\widehat{\mathbf{u}}$) from the desired value:

$$\widehat{\mathbf{u}} = \mathbf{u}^n + \Delta t \, \mathbf{rhs}^{n+1/2}. \tag{4.22}$$

A simple algebraic rearrangement of equations (4.22) and (4.21) would then give $\mathbf{f}^{n+1/2} = (\mathbf{V} - \widehat{\mathbf{u}})/\Delta t$. This forcing expression, however, is impractical as, similarly to the direct forcing, it is based on Eulerian quantities, which is not the case for $\mathbf{V}$ defined in a different position.

The suggestion of Uhlmann (2005) is therefore to compute the forcing directly on the boundary using

$$\mathbf{F}^{n+1/2}(\mathbf{X}_l) = \frac{\mathbf{V}(\mathbf{X}_l) - \widehat{\mathbf{U}}(\mathbf{X}_l)}{\Delta t}, \tag{4.23}$$

for all force nodes $\mathbf{X}_l$ on the immersed boundary. Equation (4.23), however, needs the fluid velocity at the force node $\widehat{\mathbf{U}}(\mathbf{X}_l)$ which is obtained by appropriate transfer functions. In turn, once $\mathbf{F}^{n+1/2}(\mathbf{X}_l)$ is computed from equation (4.23), the same transfer functions can be used to compute $\mathbf{f}^{n+1/2}$ on the Eulerian grid to obtain the updated velocity field from (4.20).

The sequence of steps of the method is detailed in Algorithm 4.1 and the only point to be discussed further is the structure of the transfer functions to move quantities from the boundary to the grid and vice versa.

Lagrangian to Eulerian Forcing Using Regularized δ-Functions
The first step consists of associating a discrete volume ΔV_l to each force point $\mathbf{X}_l$ such that the union of all these volumes forms a thin shell around the immersed

---

Algorithm 4.1 Sequence of steps for a Lagrangian IBM

---

1) Compute $\widehat{\mathbf{u}}$ from (4.22);

2) Transfer the Eulerian field $\widehat{\mathbf{u}}$ to $\widehat{\mathbf{U}}$ on all force points by transfer functions;

3) Compute Lagrangian IB forcing $\mathbf{F}^{n+1/2}$ on all force nodes from (4.23);

4) Transfer back the Lagrangian forcing $\mathbf{F}^{n+1/2}$ to the Eulerian field $\mathbf{f}^{n+1/2}$ using the same transfer functions as step 2;

5) Plug $\mathbf{f}^{n+1/2}$ into equation (4.20) and compute the updated Eulerian velocity field $\mathbf{u}^{n+1}$;

6) Go to step 1 to advance another Δt.

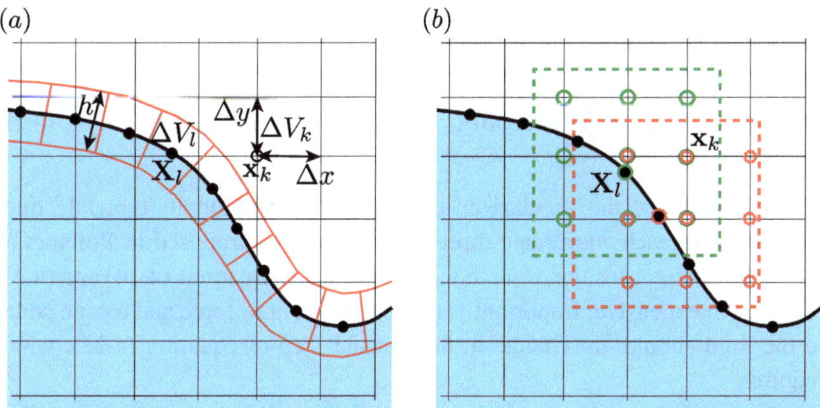

Figure 4.9 (a) Lagrangian markers distributed along the immersed boundary (black bullets) and corresponding Lagrangian volumes (red trapezoids) used for the reconstruction. (b) "Cage" of Eulerian points surrounding Lagrangian markers, used for the MLS reconstruction.

boundary of thickness about equal to one Eulerian mesh width ($h \approx \Delta x, \Delta y$ in Figure 4.9). Additionally, a volume $\Delta V_k = \Delta x \Delta y \Delta z$ is associated with the kth Eulerian node and the requirement $\Delta V_l \approx \Delta V_k$ determines the spacing between two consecutive force nodes. If the immersed object is described by a triangulated surface (as the STL or GTS standards described in Chapter 3) then, indicated by A_h the area of each triangle, the auxiliary volume thickness is such that $A_h \cdot h = \Delta V_l$. The force points $\mathbf{X}_l$ are positioned at the geometrical centroid of each triangle, whose area should be evenly distributed and proportional to the square of the local grid size.

The transfer function proposed by Uhlmann (2005) is based on the regularized δ-functions introduced in Peskin (1972a) and given by equation (4.5). The original implementation of Uhlmann (2005) was performed on a Cartesian grid with uniform spacing $\Delta = \Delta x = \Delta y = \Delta z$ to ease the interpolation procedure.

Dropping the temporal superscripts for convenience, the interpolated velocity at the Lagrangian nodes $\mathbf{X}_l$ is written as

$$\widehat{\mathbf{U}}(\mathbf{X}_l) = \sum_{k=1}^{N_e} \widehat{\mathbf{u}}(\mathbf{x}_k)\delta(\mathbf{x}_k - \mathbf{X}_l)\Delta V_k \quad \forall \; 1 \le l \le N_l, \tag{4.24}$$

where N_e is the number of Eulerian nodes, N_l is the number of Lagrangian markers and ΔV_k is the corresponding cell volume of the Eulerian grid. The continuously differentiable function δ of equation (4.5)–(4.7) yields a smooth transfer between Eulerian and Lagrangian quantities.

The IB forcing (4.23) is transferred to the Eulerian nodes with generic coordinate $\mathbf{x}_k$ using the same interpolation rule:

$$\mathbf{f}(\mathbf{x}_k) = \sum_{l=1}^{N_l} \mathbf{F}(\mathbf{X}_l)\delta(\mathbf{x}_k - \mathbf{X}_l)\Delta V_l \quad \forall \; 1 \le k \le N_e. \tag{4.25}$$

The support of the regularized delta function δ is small (typically three grid points in each coordinate direction are used, as proposed in Roma et al. (1999)), which makes the evaluation of the sums in equations (4.24) and (4.25) relatively inexpensive. Importantly, the total amount of force and torque added to the fluid should not change in the transfer step of equation (4.25), which requires

$$\sum_{k=1}^{N_e} \mathbf{f}(\mathbf{x}_k)\Delta V_k = \sum_{l=1}^{N_l} \mathbf{F}(\mathbf{X}_l)\Delta V_l,$$

$$\sum_{k=1}^{N_e} (\mathbf{x}_k - \mathbf{x}_O) \times \mathbf{f}(\mathbf{x}_k)\Delta V_k = \sum_{l=1}^{N_l} (\mathbf{X}_l - \mathbf{x}_O) \times \mathbf{F}(\mathbf{X}_l)\Delta V_l, \quad \forall \; \mathbf{x}_O. \tag{4.26}$$

As shown by Uhlmann (2005), these conservation rules hold exactly only for uniform Eulerian grids; this is the main reason why strictly uniform meshes are usually employed with this method.

Lagrangian to Eulerian Forcing Using MLS

Vanella and Balaras (2009) modified the approach proposed by Uhlmann (2005) using a versatile moving-least-squares (MLS) approximation to build the transfer functions between the Eulerian and Lagrangian grids for rigid bodies.

The procedure starts with the definition of a transfer function to approximate $\widehat{\mathbf{U}}$ for any point $\mathbf{x}$ in the neighborhood of the marker $\mathbf{X}_l$. For the ease of notation, we will consider the ith component of the vector $\widehat{\mathbf{U}}$, namely $\widehat{U}_i$, to write

$$\widehat{U}_i(\mathbf{x}) = \sum_{j=1}^{m} p_j(\mathbf{x}) a_j = \mathbf{p}^T(\mathbf{x}) \mathbf{a}, \tag{4.27}$$

with $\mathbf{p}^T$ the vector of the basis functions, of length m, and $\mathbf{a}$ a vector, also of length m, of unknown interpolation coefficients.

The choice of the basis function determines the accuracy of the MLS interpolation and it is typically taken equal to that of the spatial discretization method. For second-order-accurate methods, a linear basis function is selected with $m = 4$ and $\mathbf{p}^T(\mathbf{x}) = [1\ x\ y\ z]$ for three-dimensional problems.

If the support domain for the MLS interpolation around the force point $\mathbf{X}_l$ contains N_k Eulerian nodes $\mathbf{x}_k$ (see Figure 4.9b), the unknown interpolation coefficients $\mathbf{a}$ are obtained by minimizing the weighted L_2 norm of the difference $\widehat{U}_i - \widehat{u}_i$ for all $\mathbf{x}_k$ which, on account of (4.27), yields

$$J = \sum_{k=1}^{N_k} W(\mathbf{X}_l - \mathbf{x}_k) \left[\mathbf{p}^T(\mathbf{x}_k) \mathbf{a} - \hat{u}_i(\mathbf{x}_k) \right]. \tag{4.28}$$

The support domain is centered at $\mathbf{X}_l$ and it usually extends to $\pm r_i$, with $r_i = 1.5\Delta x_i$ so that $N_k = 27$ in three dimensions (or $N_k = 9$ for the two-dimensional sketch of Figure 4.9b).

The weight function, $W(\mathbf{X}_l - \mathbf{x}_k)$, plays an important role in the performance of the MLS interpolation: It has to be positive defined and smooth within the support domain, with a maximum for $\mathbf{X}_l - \mathbf{x}^k = 0$ and decreasing monotonically to zero outside the support domain. Let r_k be the distance of the marker $\mathbf{X}_l$ from the grid point $\mathbf{x}_k$ rescaled by the previously defined size of the support domain in the ith direction r_i:

$$r_k = \frac{|\mathbf{X}_l - \mathbf{x}_k|}{r_i}. \tag{4.29}$$

The most commonly used weight functions are: The cubic spline with second-order continuity

$$W(\mathbf{X}_l - \mathbf{x}_k) = \begin{cases} 2/3 - 4r_k^2 + 4r_k^3, & r_k \le 0.5, \\ 4/3 - 4r_k + 4r_k^2 - 4/3r_k^3 & 0.5 < r_k \le 1, \\ 0 & r_k > 1, \end{cases} \tag{4.30}$$

the quartic spline with third-order continuity

$$W(\mathbf{X}_l - \mathbf{x}_k) = \begin{cases} 1 - 6r_k^2 + 8r_k^3 - 3r_k^4, & r_k \le 1, \\ 0 & r_k > 1, \end{cases} \tag{4.31}$$

and the exponential function:

$$W(\mathbf{X}_l - \mathbf{x}_k) = \begin{cases} e^{-(r_k/\alpha)^2} & r_k \leq 1, \\ 0 & r_k > 1, \end{cases} \tag{4.32}$$

where α is a shape parameter, with lower (higher) values of α concentrating (spreading) the weight function around the sampling point $\mathbf{X}_l$. This parameter is usually set to $\alpha = [0.3 - 0.7]$ in MLS immersed boundary methods (Liu and Gu, 2005; de Tullio and Pascazio, 2016).

The optimal condition $\partial J / \partial \mathbf{a} = 0$ yields the following 4×4 linear system:

$$\mathbf{Aa} = \mathbf{B}\widehat{\mathbf{u}}_i \implies \mathbf{a} = \mathbf{A}^{-1}\mathbf{B}\widehat{\mathbf{u}}_i \tag{4.33}$$

with

$$\begin{aligned}
\mathbf{A} &= \sum_{k=1}^{N_k} W(\mathbf{X}_l - \mathbf{x}_k)\mathbf{p}(\mathbf{x}_k)\mathbf{p}^T(\mathbf{x}_k), \\
\mathbf{B} &= [W(\mathbf{X}_l - \mathbf{x}_1)\mathbf{p}(\mathbf{x}_1), \ldots, W(\mathbf{X}_l - \mathbf{x}_{N_k})\mathbf{p}(\mathbf{x}_{N_k})], \text{ and} \\
\widehat{\mathbf{u}}_i &= [\widehat{u}_i(\mathbf{x}_1) \ldots \widehat{u}_i(\mathbf{x}_{N_k})]^T.
\end{aligned} \tag{4.34}$$

Substituting equation (4.33) into equation (4.27) yields the interpolated velocity component

$$\widehat{U}_i(\mathbf{X}_l) = \mathbf{\Phi}^T(\mathbf{X}_l)\widehat{\mathbf{u}}_i = \sum_{k=1}^{N_k} \phi_k(\mathbf{X}_l)\widehat{u}_i(\mathbf{x}_k), \tag{4.35}$$

with $\mathbf{\Phi}^T(\mathbf{X}_l) = \mathbf{p}^T(\mathbf{X}_l)\mathbf{A}^{-1}\mathbf{B}$. The MLS interpolation weights ϕ_k thus allow us to evaluate the velocity components $\widehat{U}_i(\mathbf{X}_l)$ via a linear combination of the Eulerian velocities within the support domain.

Note that, as the number of nodes of the support domain N_k is larger than the number of unknown coefficients m, the interpolated velocity evaluated at the Eulerian nodes is generally different from the Eulerian velocity at the same locations, $\widehat{U}_i(\mathbf{x}_k) \neq \widehat{u}_i(\mathbf{x}_k)$; the proposed method, however, yields the minimum global mismatch between the two quantities and the weight function in (4.28) makes the difference the smallest around $\mathbf{X}_l$.

The interpolated velocities $\widehat{U}_i$ from equation (4.35) can then be used for the ith component of equation (4.23) to obtain the volume force F_i on the Lagrangian force points. In order to transfer back F_i to the Eulerian grid points, the same shape functions employed for the interpolation are used:

$$f_i(\mathbf{x}_k) = \sum_{l=1}^{N_{lk}} c_l \phi_k(\mathbf{X}_l) F_i(\mathbf{X}_l), \tag{4.36}$$

where N_{lk} is the number of force points "associated" with the Eulerian node k

(or the Lagrangian markers whose support domain contains the grid point $\mathbf{x}_k$).
The additional coefficients c_l in equation (4.36) are introduced to ensure that
the total force acting on the fluid is not altered by the transfer:

$$\sum_{k=1}^{N_e} f_i(\mathbf{x}_k)\Delta V_k = \sum_{l=1}^{N_l} F_i(\mathbf{X}_l)\Delta V_l, \tag{4.37}$$

where N_e and N_l are the total number of forced Eulerian grid points and
Lagrangian markers, and ΔV_k is the volume associated with each Eulerian point
(equal to the Eulerian cell volume). A volume $\Delta V_l = A_l h_l$ is also associated to
the marker, where A_l is the lth triangle area and h_l is its local thickness, equal
to the average mesh size at the marker location, $h_l = 1/3 \sum_{k=1}^{N_k} \phi_k(\mathbf{X}_l)(\Delta x_k +
\Delta y_k + \Delta z_k)$. Rearranging the terms yields

$$c_l = \frac{\Delta V_l}{\sum_{k=1}^{N_k} \phi_k(\mathbf{X}_l)\Delta V_k}. \tag{4.38}$$

The same relation can also be obtained by considering the balance of
Eulerian/Lagrangian forces for a single marker, $\sum_{k=1}^{N_k} c_l\phi_k(\mathbf{X}_l)F_i(\mathbf{X}_l)\Delta V_k =
F_i(\mathbf{X}_l)\Delta V_l$. The resulting transfer operators conserve momentum on both uni-
form and stretched grids, while the equivalence of total torque between the
Eulerian and Lagrangian grids

$$\sum_{k=1}^{N_e} (\mathbf{x}_k - \mathbf{x}_O) \times \mathbf{f}(\mathbf{x}_k)\Delta V_k = \sum_{l=1}^{N_l} (\mathbf{X}_l - \mathbf{x}_O) \times \mathbf{F}(\mathbf{X}_l)\Delta V_l, \quad \forall \; \mathbf{x}_O, \tag{4.39}$$

are, once again, guaranteed only for uniform grids, however, with minimal errors for smoothly varying grids with low stretching (Vanella and Balaras, 2009).
In order to correctly impose the boundary condition at the walls, a sufficient
resolution of the immersed surface is needed since the distance between adjacent Lagrangian markers has to be comparable with the local Eulerian spacing.
A Lagrangian spacing of about 0.7 of the Eulerian spacing has been found
to be an optimal choice, also considering the case of triangles stretching for
deformable bodies (de Tullio and Pascazio, 2016).

4.3 Numerical Treatments Inside Rigid Bodies

The discussion of Sections 4.1 and 4.2 was focused on the treatment of the
grid nodes next to the immersed boundary, or interface points, in order to
enforce a boundary condition at the wet surface. If the latter is closed, however,
there exists an inner volume whose nodes, or internal points, can be treated
in different ways depending on the flow problem and the immersed boundary
implementation.

For example, a vesicle containing a liquid and advected by another fluid carrier will have both an internal flow and an external flow whose combined action will determine the instantaneous position and shape of the membranal boundary; in problems like these, the governing equations must be solved inside and outside the interface and only the nodes closest to the boundary have to be modified by some IB forcing.

On the other hand, if the immersed body is rigid, the instantaneous velocity of any of its points $\mathbf{x}_k$ is determined by $\mathbf{u}(\mathbf{x}_k) = \mathbf{u}_G + \omega \times (\mathbf{x}_k - \mathbf{x}_G)$ with the velocity of the center of mass $\mathbf{u}_G = \mathbf{u}(\mathbf{x}_G)$ and the angular velocity ω, computed via the Newton–Euler equations (see Chapter 7). In this case, the inner volume contains *dead* nodes for the problem dynamics and their treatment is dictated mostly by numerical considerations, and several alternatives are possible.

A first approach consists of leaving the interior of the body free to develop an internal flow without imposing any conditions except for the IB forcing at the boundary points. In this case, artificial flow patterns, driven by the shear stresses at the wet side of the boundary, may form in the inner body volume. These patterns clearly do not have a physical meaning; this is the case of the two counter-rotating vortices appearing inside the solid cylinder shown in Figure 4.10a.

Another possibility is to apply the IB forcing also at the internal nodes to impose the solid body velocity $\mathbf{u}(\mathbf{x}_k)$ (which reduces to $\mathbf{u}(\mathbf{x}_k) = \mathbf{0}$ for a fixed body) with the pressure that adjusts accordingly. Of course, in this case, any internal flows are prevented, although the external flow is unchanged (see Figure 4.10b). As an additional constraint for non-accelerating rigid bodies, since the inner velocity field is constant in space and time, the internal pressure can be set to zero, as has been done in the case in Figure 4.10c.

A final possibility is to add the forcing at the first internal node using the same rule as the IB forcing at the first external point. For the velocity, this method basically mirrors the velocity at the first interior node (point M in Figure 4.10d) to the one at the first external point (point A in the same figure). This approach combines the external forcing with a ghost cell. Hence, the boundary conditions are imposed on two layers of nodes rather than a single one. This reinforced IB treatment might be advantageous if the resolutions of Eulerian mesh and surface tessellation are too different, and *orphan* nodes, that is, interface grid points which do not enforce a boundary condition, may arise. In this case, the double layer of forced nodes across the immersed boundary avoids undesired surface transpiration, although simulations with orphan nodes should always be avoided.

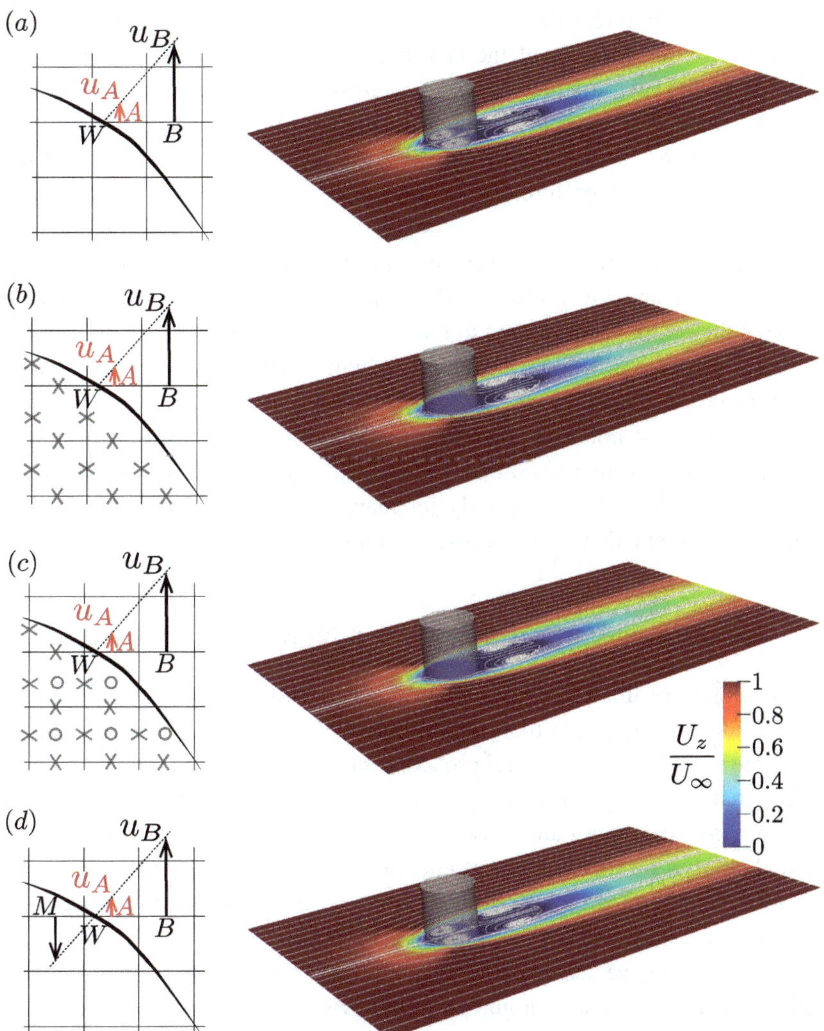

Figure 4.10 Flow around a three-dimensional circular cylinder at Reynolds number $Re = 40$ based on the incoming velocity and cylinder diameter. The streamwise velocity field with superimposed streamlines is shown for the symmetry plane of the cylinder. Four different treatments for the internal nodes are considered: (a) no internal treatment, (b) blanking of the internal velocity, (c) blanking of the internal velocity and pressure and (d) mirroring of the velocity at the first internal point.

The panels of Figure 4.10 clearly show that, despite the very different internal flow patterns, the features of the external flow are the same. No significant variation of the recirculation bubble dimension and strength, nor changes in the drag coefficient have been observed for any of the methods presented. Essentially, there is no influence on the accuracy or on the efficiency of the numerical scheme; therefore, depending on the particular flow, the easiest method should be used.

It is important to note, however, that the internal flow details are very dependent on the specific geometry; thus, should the highest flow velocity be localized therein, it is convenient to force the inner volume so as to prevent the formation of inner flow, not to uselessly penalize the integration time step of the whole computation.

As an aside, we note that whenever the velocity field is known inside the body, enforcing its distribution by IB forcings can mitigate problems in the boundary reconstruction, especially for marginally refined meshes, and thus improve the satisfaction of the boundary conditions.

4.4 Immersed Boundaries and Numerical Methods

The beginning of this chapter has been devoted to the analysis of various expressions for $\mathbf{f}$ in equations (4.1) employed to enforce specific conditions on an immersed boundary; then several procedures to reconstruct the solution along a surface that is not aligned with any coordinate lines have been discussed in detail, stressing advantages and drawbacks. Although any of the aforementioned methods hinges on the solution of equations (4.1), we have been very vague about how this is effectively carried out and whether or not the introduction of the forcing $\mathbf{f}$ interferes with the solution procedure.

In the following we describe, among many possibilities, two popular solution methods to give the reader a guideline to devise a practical solution strategy and also to underline how some IBM issues depend on the coupling with a specific solution algorithm rather than on the method itself.

4.4.1 The Fractional-Step Method

This method is one of the most popular for simulations of incompressible viscous flows and it has been proposed and validated in the seminal papers by Kim and Moin (1985) and Rai and Moin (1991); here we report the details of the various steps since it is the same algorithm implemented in the flow solver employed for the examples in Chapter 10, whose source files are made available with this book.

Once again, we start from the incompressible Navier–Stokes equations in non-dimensional form,

$$\nabla \cdot \mathbf{u} = 0,$$

$$\frac{\partial \mathbf{u}}{\partial t} + \nabla \cdot (\mathbf{uu}) = -\nabla p + \frac{1}{Re}\nabla^2 \mathbf{u} + \mathbf{f}, \tag{4.40}$$

whose derivation was detailed in Chapter 2. The goal is to devise a numerical algorithm to advance in time velocity $\mathbf{u}$ and pressure p fields starting from assigned initial and boundary conditions.

The time-discretization of the momentum equation, using an explicit Adams–Bashforth method for the nonlinear convective term and an implicit Crank–Nicolson method for the viscous ones, yields

$$\frac{\mathbf{u}^{n+1}-\mathbf{u}^n}{\Delta t}+\left[\frac{3}{2}\nabla \cdot (\mathbf{uu})^n - \frac{1}{2}\nabla \cdot (\mathbf{uu})^{n-1}\right]=-\nabla p^{n+1}+\frac{1}{2Re}\nabla^2(\mathbf{u}^{n+1}+\mathbf{u}^n)+\mathbf{f}^{n+1/2}, \tag{4.41}$$

where the superscripts indicate the time-discretized level at which quantities are evaluated.

A major problem in the solution of this equation is that it contains both unknown fields, $\mathbf{u}^{n+1}$ and p^{n+1}, and it is the only one containing a time derivative. The reason is that, in incompressible flows, pressure is not a dynamic variable while it has rather the role of projecting any velocity distributions onto a divergence-free space in which mass conservation is automatically satisfied, $\nabla \cdot \mathbf{u}^{n+1} = 0$. As an aside, we note that this is the reason why only the updated pressure p^{n+1} is used in (4.41), rather than the average between the time levels n and $n + 1$ as for velocity.

Since equation (4.41) can not be solved simultaneously for $\mathbf{u}^{n+1}$ and p^{n+1}, a fractional step method is used in which a provisional (non-solenoidal) velocity $\widehat{\mathbf{u}}$ is obtained from the time-discrete equation with the previous pressure:

$$\frac{\widehat{\mathbf{u}}-\mathbf{u}^n}{\Delta t} = -\left[\frac{3}{2}\nabla \cdot (\mathbf{uu})^n - \frac{1}{2}\nabla \cdot (\mathbf{uu})^{n-1}\right] - \nabla p^n + \frac{1}{2Re}\nabla^2(\widehat{\mathbf{u}} + \mathbf{u}^n) + \mathbf{f}^{n+1/2}. \tag{4.42}$$

The presence of the unknown $\widehat{\mathbf{u}}$ in the Laplacian operator on the right-hand side (caused by the implicit treatment of the viscous terms) would entail the inversion of a sparse matrix that, for the number of nodes of a typical Eulerian grid, requires an unbearable computational burden. For second-order finite difference discretizations, however, this can be avoided by using the approximate factorization technique of Beam and Warming (1976), which yields the product of three tridiagonal matrices efficiently solved via Thomas' algorithm. The IB forcing $\mathbf{f}^{n+1/2}$ can be defined either implicitly, using $\widehat{\mathbf{u}}$ and $\mathbf{u}^n$, or explicitly with

$\mathbf{u}^n$ and $\mathbf{u}^{n-1}$, as in the numerical code we use in this book (see Chapter 10).

If the IB forcing is defined explicitly, $\mathbf{f}$ can be removed from the equation for the provisional velocity (4.42) and applied in an additional step to the intermediate velocity $\mathbf{u}^*$ as follows:

$$\frac{\mathbf{u}^* - \widehat{\mathbf{u}}}{\Delta t} = \mathbf{f}^{(n+1/2)} \qquad \Rightarrow \qquad \mathbf{u}^* = \widehat{\mathbf{u}} + \Delta t\, \mathbf{f}^{(n+1/2)}. \tag{4.43}$$

We anticipate that in the numerical code distributed with the book (in Chapter 10 there is a link for the download), the *Eulerian* IB forcing is applied in equation (4.42), whereas the *Lagrangian* one is applied through the intermediate step (4.43).

The non-solenoidal velocity $\mathbf{u}^*$ (or directly $\widehat{\mathbf{u}}$ if the IB forcing is included implicitly in the first substep (4.42)), is projected onto a divergence-free space by applying the pressure correction. To this aim, it is observed that the only difference between (4.42) and (4.41) is the pressure gradient ∇p^n in the former rather than the correct one ∇p^{n+1}; it is therefore expected that the difference between $\widehat{\mathbf{u}}$ and $\mathbf{u}^{n+1}$ must be in the form of a gradient of a scalar quantity, which can be written as

$$\frac{\mathbf{u}^{n+1} - \mathbf{u}^*}{\Delta t} = -\nabla\Phi \qquad \Rightarrow \qquad \mathbf{u}^{n+1} = \mathbf{u}^* - \Delta t \nabla\Phi. \tag{4.44}$$

The scalar field Φ is obtained by taking the divergence of this correction and imposing the solenoidal character of the updated velocity field $\mathbf{u}^{n+1}$:

$$\nabla^2\Phi = \frac{\nabla \cdot \mathbf{u}^*}{\Delta t}, \tag{4.45}$$

with $\nabla\Phi \cdot \mathbf{n} = 0$ at the boundaries of the computational domain or periodic conditions in the periodic directions. Note that in the case of solid boundaries, the homogeneous Neumann condition ensures that the corrected velocity remains tangent to the wall. The elliptic equation (4.45) is solved using iterative methods or direct methods for Poisson problems, which are based on fast Fourier transformations (FFTs) (Orlandi, 2012).

The updated pressure p^{n+1} is then determined by a complementary relation defined in such a way that, when added to equation (4.42), it returns equation (4.41):

$$\frac{\mathbf{u}^{n+1} - \mathbf{u}^*}{\Delta t} = -\nabla p^{n+1} + \nabla p^n + \frac{1}{2Re}\nabla^2(\mathbf{u}^{n+1} - \mathbf{u}^*), \tag{4.46}$$

with the only unknown being p^{n+1}. Inserting equation (4.44) into (4.46) and using the commutativity of Laplacian and gradient operators, we obtain the result

$$p^{n+1} = p^n + \Phi - \frac{\Delta t}{2Re} \nabla^2 \Phi. \tag{4.47}$$

4.4.2 Pseudo-Compressibility Method

In Section 4.4.1, we have seen that satisfying the flow incompressibility $\nabla \cdot \mathbf{u} = 0$ brings us to the elliptic equation (4.45) which, for large Eulerian grids, is the computational bottleneck of the algorithm. This observation motivates a change in the governing equations in order to avoid this issue: This is the pseudo-compressibility method. In this context, incompressibility is relaxed by adding a compressible part to the continuity equation to yield the system

$$\frac{1}{\beta} \frac{\partial p}{\partial t} + \nabla \cdot \mathbf{u} = 0,$$
$$\frac{\partial \mathbf{u}}{\partial t} + \nabla \cdot (\mathbf{uu}) = -\nabla p + \frac{1}{Re} \nabla^2 \mathbf{u} + \mathbf{f}, \tag{4.48}$$

which can be integrated by a time-marching scheme. Here $\beta > 0$ is the pseudo-compressibility factor; it does not have a physical meaning and it must be provided as a free parameter based only on numerical considerations. Needless to say, while the system (4.40) is derived on physical grounds, the modified counterpart (4.48) does not benefit from the same property and its transient solution does not reproduce any physical systems.

However, should the flow attain a steady state, all time-derivative terms would vanish and the solutions of (4.40) and (4.48) would coincide, with the former not requiring the numerical solution of an elliptic equation.

In fact, considering the following discretization:

$$\frac{1}{\beta} \frac{p^{n+1} - p^n}{\Delta t} + \nabla \cdot \mathbf{u}^{n+1} = 0,$$
$$\frac{\mathbf{u}^{n+1} - \mathbf{u}^n}{\Delta t} + \nabla \cdot (\mathbf{uu})^n = -\nabla p^{n+1} + \frac{1}{Re} \nabla^2 \mathbf{u}^{n+1} + \mathbf{f}^{n+1/2}, \tag{4.49}$$

its first-order accuracy in time is inconsequential when a steady-state solution is achieved and it can be obtained by a simple integration scheme not requiring large computational resources.

This method has also been extended to unsteady problems by introducing an auxiliary fictitious time τ over which the solution evolves for every physical time step Δt to obtain a converged instantaneous configuration.

The new set of equations reads

$$\frac{1}{\beta}\frac{\partial p}{\partial \tau} + \nabla \cdot \mathbf{u} = 0,$$

$$\frac{\partial \mathbf{u}}{\partial \tau} + \frac{\partial \mathbf{u}}{\partial t} + \nabla \cdot (\mathbf{u}\mathbf{u}) = -\nabla p + \frac{1}{Re}\nabla^2 \mathbf{u} + \mathbf{f},$$

$$(4.50)$$

with the first unsteady term in the momentum equation introducing a numerical transient in the artificial pseudo-time τ (Louda et al., 2008).

In this method, the artificial speed of sound β is a key parameter whose choice is a tradeoff between computational efficiency and mathematical accuracy. Increasing the value of β improves accuracy, as equations (4.50) approach the incompressible limit, but simultaneously the computational effort increases owing to restrictions on the time step. In contrast, small β allows for faster integration at the cost of a larger error in the velocity divergence.

In summary, the absence of a Poisson equation comes at the cost of a pseudo-transient at every time step and a user-defined parameter (β), which controls the error in the velocity incompressibility (Chung, 2002). Typical values of β are in the range 0.1–10 (Kwak et al., 1986), depending on the particular problem. It appears, however, that for strongly unsteady turbulent flows, the overall computational load becomes too large to obtain a solution as a direct numerical simulation and the aforementioned scheme is restricted to equations with turbulence models (see Chapter 8) in which the fast flow dynamics is filtered out and the fields can be advanced in time using large steps (Δt).

4.5 Caveats of the Immersed Boundaries

4.5.1 Enforcing Local Mass Conservation

The main reason for the ease and computational efficiency of IBMs, at least in the versions discussed so far, is that all the alterations of the method are made to the momentum equation, while mass conservation $\nabla \cdot \mathbf{u} = 0$ is not modified.

However, the immersed boundary procedure generates two regions in which velocity components near the IB are determined separately on each side and the resulting velocity field turns out to be not divergence-free across the boundary. On the other hand, if this velocity is projected onto a solenoidal field (see equation (4.44)), mass conservation must also be locally satisfied for the cells crossed by the immersed boundary, although the correction can perturb the correct enforcement of the boundary condition. For example, in the case of a solid body, the external solution should be totally decoupled from that inside, but the flow field in the near-wall region is affected by the velocity in the solid

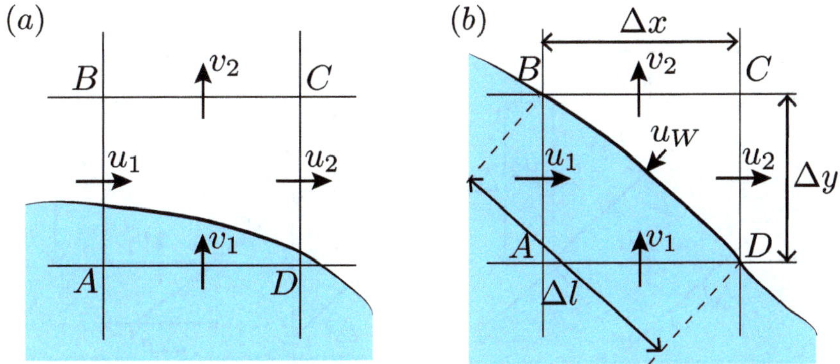

Figure 4.11 (a) Schematic of a solid body immersed in a staggered grid. The velocity components of the external grid points (u_1, u_2 and v_2) are imposed by the IB forcing. (b) Treatment of the divergence of the velocity for a control volume crossed by the IB.

since the discrete velocity divergence is computed at every cell regardless of the presence of the IB.

Figure 4.11a shows an example of a solid body immersed in a staggered two-dimensional grid where the vertical (horizontal) velocity is defined at the center of the horizontal (vertical) face. For a boundary reconstruction scheme based on external grid points (see Section 4.2.1), the velocity components therein u_1, u_2 and v_2 are enforced by the IB forcing (see equation (4.17)). According to the possible treatments of the solid region discussed in Section 4.3, the inner velocity component, v_1, can be either set to zero or mirrored by a linear extrapolation of the corresponding external velocity. In both cases, conservation of mass through the control volume $ABCD$ is not fulfilled as all momentum fluxes are defined by the IB forcing without locally accounting for the mass conservation. The latter issue is mitigated when IBMs are combined with a fractional-step method, thanks to the final projection step of equation (4.44), although velocities are altered across the IB, as explained in Section 4.5.2. Even if any treatment were applied inside the solid, and the velocity component v_1 were determined by the discretized momentum equation, more than a momentum flux in the control volume $ABCD$ is computed by IB forcings and mass conservation is not satisfied locally.

An additional problem is that mass conservation may not be satisfied simultaneously in the mesh cell crossed by the immersed body and in the virtual cell obtained from cutting the previous cells with the IB. In the example of Figure 4.11b, these two control volumes correspond, respectively, to $ABCD$ and BCD, which entail the discrete continuity equations

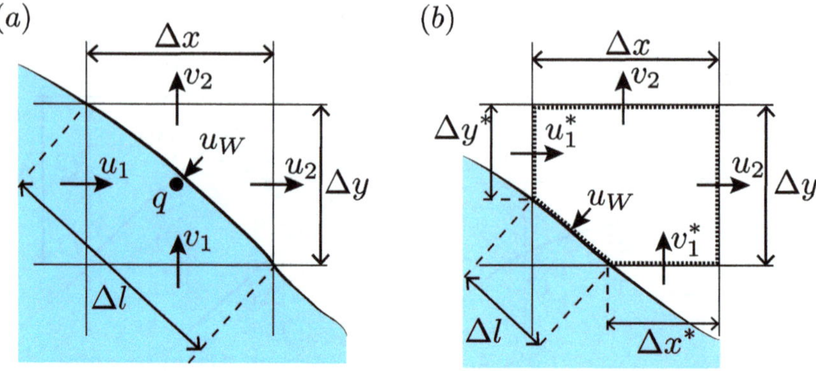

Figure 4.12 (a) Schematic of the source/sink method of Kim et al. (2001). (b) Treatment of the divergence of the velocity for a control volume crossed by the IB according to the cut-cell method.

$$u_1\Delta y + v_1\Delta x - u_2\Delta y - v_2\Delta x = 0, \tag{4.51}$$

and

$$-u_W\Delta l - u_2\Delta y - v_2\Delta x = 0, \tag{4.52}$$

with u_W the velocity of the boundary, which is zero for a fixed body.

The problem has been addressed by Kim et al. (2001), who added a system of mass source/sink to the continuity equation to satisfy mass conservation for fluid cells containing the immersed boundary. In particular, a source is introduced, as in Figure 4.12a, at the center of the cell containing both body and fluid. The continuity equation (4.51) then becomes

$$u_1\Delta y + v_1\Delta x - u_2\Delta y - v_2\Delta x + q\Delta x\Delta y = 0. \tag{4.53}$$

From equations (4.52) and (4.53) the mass source q for a fixed body ($u_W = 0$) is obtained as

$$q = -\frac{u_1}{\Delta x} - \frac{v_1}{\Delta y}. \tag{4.54}$$

This additional degree of freedom, hence, allows us to satisfy both equations (4.52) and (4.53). However, this approach is formulated for linear stepwise geometries whose boundary is aligned with the grid points and the resulting velocity field will not be divergence-free, although the additional mass source or sink terms satisfy global mass conservation.

Another approach can be implemented in the finite-volume formulation, without resorting to alterations of the continuity equation. Using the cut-cell method (Ye et al., 1999; Kirkpatrick et al., 2003), the solid region can be decoupled from the fluid counterpart by building new computational cells formed by the IB and existing cell faces. Figure 4.12b shows a schematic of a Cartesian grid with an IB separating a solid from a fluid. In this method, cells cut by the IB whose cell center lies in the fluid are reshaped by discarding their solid portion. Pieces of cut cells whose centers lie in the solid are absorbed by neighboring solid cells. This procedure results in the formation of modified control volumes such as the trapezoidal one in Figure 4.12b, where the continuity equation reads

$$\int\int_{\Delta V_{x,y}} \nabla \cdot \mathbf{u}\, \mathrm{d}x\mathrm{d}y = u_2\Delta y + v_2\Delta x - u_1^*\Delta y^* - v_1^*\Delta x^* + u_W\Delta l + O\left(\Delta x^3\right). \quad (4.55)$$

Here, the velocity component at the center of the face is located in either the fluid or solid region and the superscript $*$ denotes newly defined velocity values, which are necessary for computing the divergence accurately. Interpolation is used to compute u_1^* and v_1^*. It should be noted that the finite-volume discretization for the cut cells introduces substantial complexity in the treatment of the immersed bodies when considering the variety of possibilities of boundary–cell intersections. With the proposed changes, the method basically resorts to a body-fitted treatment with the boundary computational stencils requiring the modification of the solver with a potential negative impact on the efficiency of the code.

4.5.2 Perturbation of the IB Forcing within a Fractional-Step Procedure

A subtle, though relevant, issue of the IB momentum forcing arises when such methods are combined with Navier–Stokes solvers based on fractional-step schemes. In fact, as anticipated in Section 4.4.1, the no-slip condition on the velocity field is enforced on the provisional velocity $\mathbf{u}^*$ (see equation (4.43)), which is perturbed during the projection (4.44). While this correction step is needed to have a mass-conserving field $\mathbf{u}^{n+1}$, it alters the velocity field $\mathbf{u}^*$, which fulfills the condition imposed through the IBM forcing at the immersed boundary. As a consequence, the final velocity field $\mathbf{u}^{n+1}$ may not satisfy exactly the boundary condition for which the IBM has been devised at the body surface. A first observation is that in the fractional-step method of Section 4.4.1 the velocity correction is of order $||\mathbf{u}^{n+1} - \mathbf{u}^*|| = O(\Delta t^2)$; thus, not only is the numerical accuracy of the scheme preserved, but the error introduced in the projection step may be arbitrarily reduced by decreasing the time step of the simulation (Guy and Hartenstine, 2010), although at the cost of increasing the time to solution proportionally.

As an alternative, the projection error can be reduced to round-off values by iterating between equations (4.43)–(4.45) of the fractional-step method, thus minimizing a potential flux through the boundary and satisfying the imposed boundary condition within a given tolerance. Note, however, that for large enough meshes, the solution of equation (4.45) becomes the most time-consuming part of the integration and a judicious evaluation of the number of iterations is necessary in order to define the integration strategy.

Such an issue on the perturbation of the IB forcing owing to the projection step is naturally circumvented by the method of Kim et al. (2001) (see Section 4.5.1), as the set of mass sources/sinks used to satisfy the continuity equation cancels the correction of equation (4.44) at the forced grid points.

A final approach to preventing the projection error is to treat the IB forcing as a Lagrange multiplier, determined to enforce the no-slip condition, while introducing appropriate regularization and interpolation operators (Glowinski et al., 1998; Taira and Colonius, 2007). The advantage of such immersed boundary projection methods (IBPM) is that continuity and no-slip conditions are satisfied implicitly by solving a modified Poisson equation (through the conjugate-gradient solver) where both IB force and pressure are unknown. The disadvantage in this case is that the simple and regular structure of the Poisson solver is spoiled, and we have to resort to more time-consuming iterative procedures.

Before concluding this section, we wish to stress that the aforementioned problem is not intrinsic to immersed boundary techniques but rather to their coupling to specific methods for the solution of Navier–Stokes equations. If the flow incompressibility is enforced by an iterative procedure (like in the pseudo-compressibility method of Section 4.4.2) or IBMs are used for compressible flows, the IB forcing **f** yields directly the correct boundary condition over the immersed body.

4.5.3 Limitations of the Lagrangian IBMs

In the explicit Lagrangian forcing methods described in Section 4.2.2, the IB forcing is evaluated at the Lagrangian level and then distributed to the surrounding fluid nodes using the spreading operator. This method, however, may produce a transpiration error at the wet boundary, especially in unsteady flows with a large pressure jump across the surface. The error in the no-slip condition has a twofold reason. Let us consider the IB-MLS case in Figure 4.13, with the Lagrangian markers $\mathbf{X}_l$ and $\mathbf{X}_m$ having support domains S_l and S_m. There exists an overlap region C_{lm} where the forcing is spread to the Eulerian

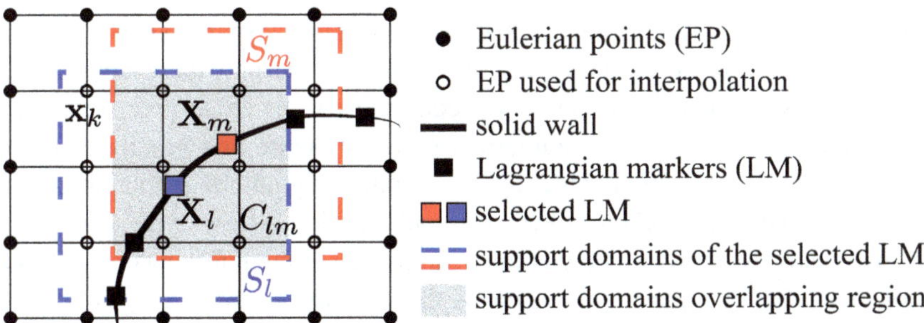

Figure 4.13 Two-dimensional sketch of the Eulerian support domains of two neighboring Lagrangian markers. The shaded area is the region of overlap of the supporting domains for the highlighted markers.

grid points $\mathbf{x}_k$ from both markers. Hence, the IB force $\mathbf{f}(\mathbf{x}_k)$ imposed by one marker may be perturbed by one of the neighboring markers. Secondly, the spreading operator is not the inverse of the interpolator: Therefore, the IB force f_i, obtained from the spreading of F_i, if interpolated back to the Lagrangian markers, does not reproduce exactly the source F_i. The latter can be easily seen if we focus on a single Lagrangian marker by considering the composition of the spreading and the interpolation operator

$$F_i\, c_l \sum_{k \in S_l} \phi_k^2(\mathbf{X}_l) \neq F_i(\mathbf{X}_l), \tag{4.56}$$

which takes a force Lagrangian field F_i, spreads it onto the Eulerian grid points and finally interpolates it back to the marker. By applying the interpolator operator to equation (4.43), we obtain the result that the discrepancy between the left- and right-hand side is proportional to a penetration velocity (Gsell and Favier, 2021).

Therefore, although being numerically efficient, Lagrangian methods have been observed not to spread correctly the IB forcing required to satisfy the boundary condition from the Lagrangian markers to the Eulerian grid. This is because of both the nonreciprocity of the spreading and interpolation operators and the overlap of the Eulerian point clouds corresponding to different Lagrangian markers.

Several amendments have been proposed to circumvent this problem, such as the implicit treatment of forcing, in which the immersed boundary forces are obtained by the solution of a linear system (Su et al., 2007); or projection approaches (Taira and Colonius, 2007), whose forcing is regarded as a Lagrange multiplier for the boundary condition constraint. Despite these approaches allowing us to impose the boundary conditions to machine precision, they require deep changes to the numerical solver structure and greatly increase the computational costs. Eventually, even employing ad hoc preconditioners, three-dimensional simulations become unaffordable.

Alternatively, explicit methods can be improved through forcing corrections. For example, a simple improvement is the introduction of forcing iterations (Kempe and Fröhlich, 2012), which increases the precision of the method with minimal modifications to its numerical implementation and a tolerable increase of computational costs. However, it has been noted that, even if decreasing the overall boundary error, the iterative approach fails to reduce it beyond a problem-dependent threshold. In fact, if the number of iterations is increased, an error saturation is always observed. In Gsell and Favier (2021) and Chen et al. (2023), a global correction factor depending on the shape of the transfer functions is introduced to correct the IB forces evaluated on the markers, whereas Vagnoli et al. (2025) defined a local and explicit correction that also adapts dynamically to the instantaneous flow: this approach is capable of reducing the error at the wet boundary without any forcing iteration. Interestingly, in Yildiran et al. (2024), the pressure equation is modified in order to explicitly impose the pressure boundary condition, which in turn reduces the boundary error when there are large pressure gradients through the walls.

Another drawback of Lagrangian IBMs comes from the need to have a fine distribution of the Lagrangian markers, hence with a surface triangulation comparable to the local Eulerian spacing (we recall that a Lagrangian spacing of about 0.7 of the Eulerian one is recommended). An immediate consequence of this constraint is that as the Eulerian grid is refined, the Lagrangian resolution must be increased as well, thus entailing geometry remeshing for every change of Eulerian resolution. This tedious problem may be avoided by using an adaptive Lagrangian mesh refinement procedure where the initial triangular mesh is automatically subdivided into virtual subtriangles (the "tiles") until each one gets smaller than the local Eulerian grid size, thus avoiding "holes" in the interfacial boundary condition. In this way, the immersed boundaries can be discretized independently of the Eulerian mesh, and each triangle is successively refined until the Lagrangian resolution of the tiled surface is sufficiently fine. The tiling procedure can be run either once at the beginning of the simulation or dynamically at each time step according to the instantaneous tissue deformation that changes the local ratio between Lagrangian and Eulerian grids. As also discussed in Viola et al. (2020), the advantage of an adaptive refinement of the Lagrangian triangulation for fluid–structure interaction is

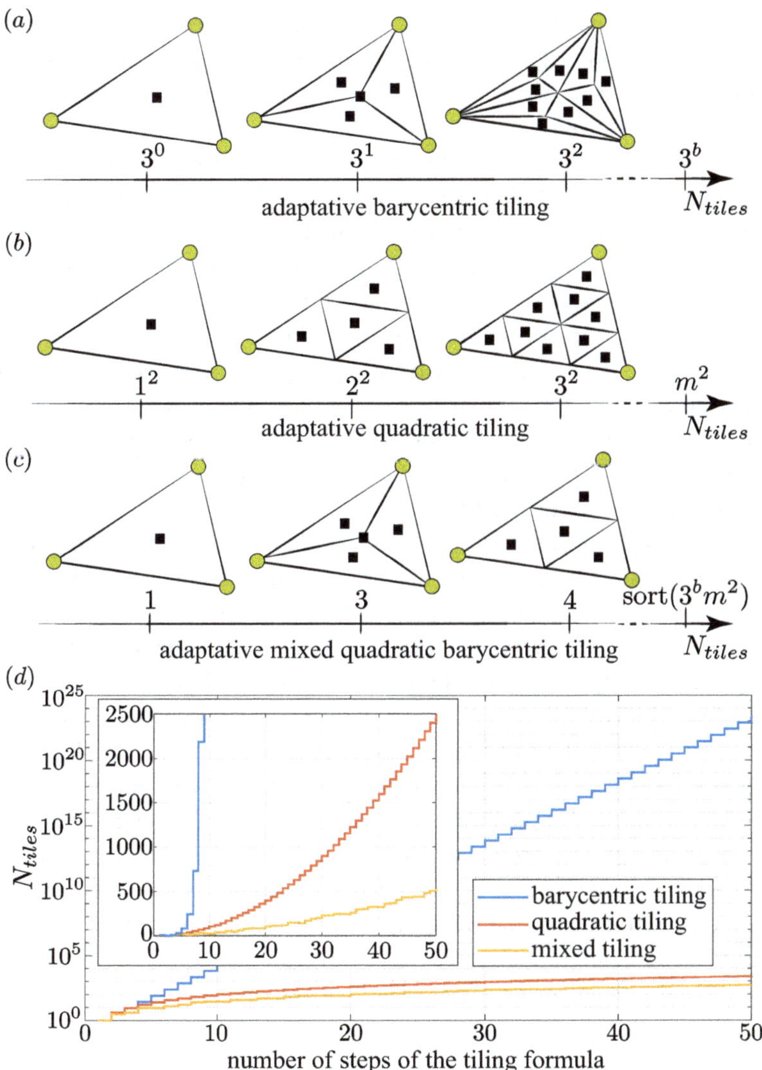

Figure 4.14 Possible adaptive tiling strategies. (a) Barycentric, (b) quadratic and (c) mixed quadratic/barycentric formula, along with the corresponding (d) number of tiles as a function of the number of steps of the formula. A slower growth in the last panel corresponds to a better control of the Lagrangian resolution through the adaptive tiling.

twofold: On one hand, it allows the use of the same base triangulation regardless of the Eulerian grid and, on the other, the number of Lagrangian nodes used to deform the immersed body can be reduced, provided that the structural loads are accurately resolved.

A possible tiling approach is the barycentric adaptive rule (Spandan et al., 2018) consisting of splitting the triangles into three parts according to the medians passing through the centroid. Higher Lagrangian resolution can thus be obtained by successive splitting of the subtriangles according to the same procedure (Figure 4.14a). The barycentric tiling rule, however, has two main drawbacks: The shape of the tiles changes with respect to the initial triangle and it gets more skewed as successive tiling steps are made (see Figure 4.14a). Furthermore, the number of tiles increases exponentially, $N_{tiles} = 3^b$, with the tiling step b (Figure 4.14d) and this limits the flexibility of the procedure as the Lagrangian resolution can only be increased by powers of 3, as does the computational cost of the IB-MLS.

These shortcomings are mitigated by an adaptive quadratic tiling (Beeson, 2012; Viola et al., 2023a) where triangles are tiled by tracing a set of $m-1$ (with $m = 1, 2, 3, \ldots$) equispaced lines parallel to each triangle edge intersecting the other two. According to Talete's theorem, all tiles are similar to the original triangle and, consequently, are similar to each other. Moreover, the number of tiles grows algebraically, $N_{tiles} = m^2$, with the tiling steps m, thus allowing a greater control of the Lagrangian resolution with respect to the barycentric tiling rule. This last aspect can be further improved by combining the quadratic tiling rule with the barycentric one. Specifically, both strategies are used to split the triangles with the number of barycentric and quadratic steps (b and m, respectively), which can be varied independently so as to obtain the desired Lagrangian refinement. The resulting "mixed" rule hence provides a larger space of tiling configurations (Figure 4.14c) with the number of tiles growing slower than for the other strategies (Figure 4.14d).

5

Moving Boundaries

All the discussion in Chapter 4 was based on the assumption that the immersed geometry is fixed with respect to the grid and all the steps required by IBMs, such as cell tagging or distance calculation for boundary reconstruction, must be performed only once at the beginning of the simulation. However, in practical applications, dealing with surfaces in (relative) motion is the norm rather than an exception, and all IBMs require substantial algorithmic adjustments in order to retain the simulation feasibility.

It is important to point out that, even accounting for this overhead, using IBMs is still more efficient than standard isogeometric methods based on body-fitted meshes; in fact, the latter schemes also pose difficult challenges and the scenario can be easily understood considering the cartoon of Figure 5.1. Here

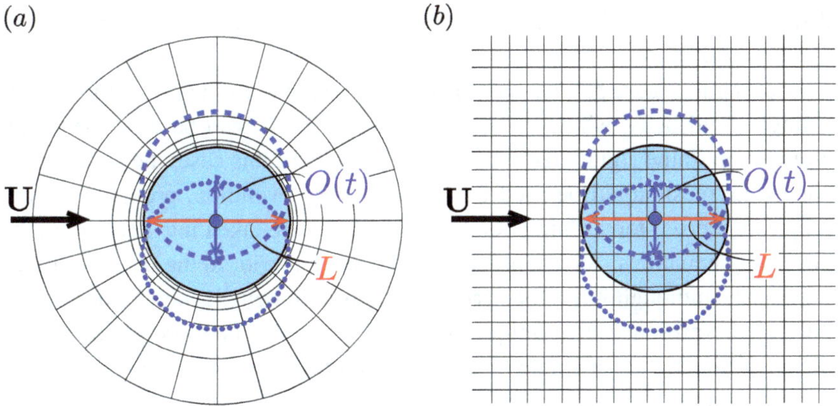

Figure 5.1 Sketch of a two-dimensional section through a sphere, subjected to a uniform flow with horizontal velocity **U**, whose position of the geometrical center is time-dependent through a given function $O(t)$: (a) discretization by body-fitted coordinates; (b) discretization by a Cartesian mesh to be employed with an IBM.

87

we consider a uniform horizontal fluid flow and a spherical solid body whose center moves vertically in time. Given the relatively simple configuration, from the body kinematics it is easy and quick to build a body-fitted spherical mesh at any instant, although for really complex geometries this part could become a time-consuming bottleneck. Furthermore, any new mesh "inherits" the flow field from the previous one; thus we must resort to interpolations, and they do not ensure the fulfillment of local and global properties, like mass or energy conservation (in addition to using a lot of computational power).

Indeed, if the boundary undergoes only small displacements the actual mesh can be obtained by deforming and stretching the previous one, though this procedure has a significant computational cost; moreover, similarly to the remeshing technique, the flow solution has to be projected or interpolated between the two grids, which entails additional overhead to the total computational time.

On the other hand, for problems with large displacements, grid deformation becomes impractical as the tiny near-wall cells tend to become extremely skewed or even to develop negative volumes. This is clearly unphysical and numerically unacceptable if the equations integration has to be kept stable and robust.

The same problem tackled by immersed boundary methods is sketched in Figure 5.1b showing that, regardless of the instantaneous body position (and its geometrical complexity), the same stationary non-body-conforming Cartesian grid can be used so that efficient numerical solvers and parallelization strategies for orthogonal structured grids can be directly applied to solve complex flow problems. IBMs offer the advantage of being applicable to both small and large displacements, as well as deforming boundaries that need determining in both location and shape.

Looking at Figure 5.1, we can further observe that, as the problem involves a single rigid body, whatever its kinematics, the flow can be described from a reference frame rigidly fixed to the body itself. In this case, the same body-conformal mesh could be used for all the system evolution, provided the appropriate inertial forces and boundary conditions are taken into account in the governing equations. This approach is certainly superior to the ones discussed previously, although it has the serious limitation of being generally applicable only in the presence of a unique body as, when two or more of them have different kinematics, a single change of reference frame can not cancel their relative motion.[1]

As an example, this is the case of the arrangement in Figure 5.2 where the same sphere as before is moving vertically relative to a flat plate at rest. It can be

---

[1] In this discussion, we have deliberately left out more sophisticated techniques, like multiblock splitting, sliding meshes or overset grids, usually aimed at industrial applications in which obtaining quick solutions, possibly with coarse-grained turbulence models, is more important than accuracy.

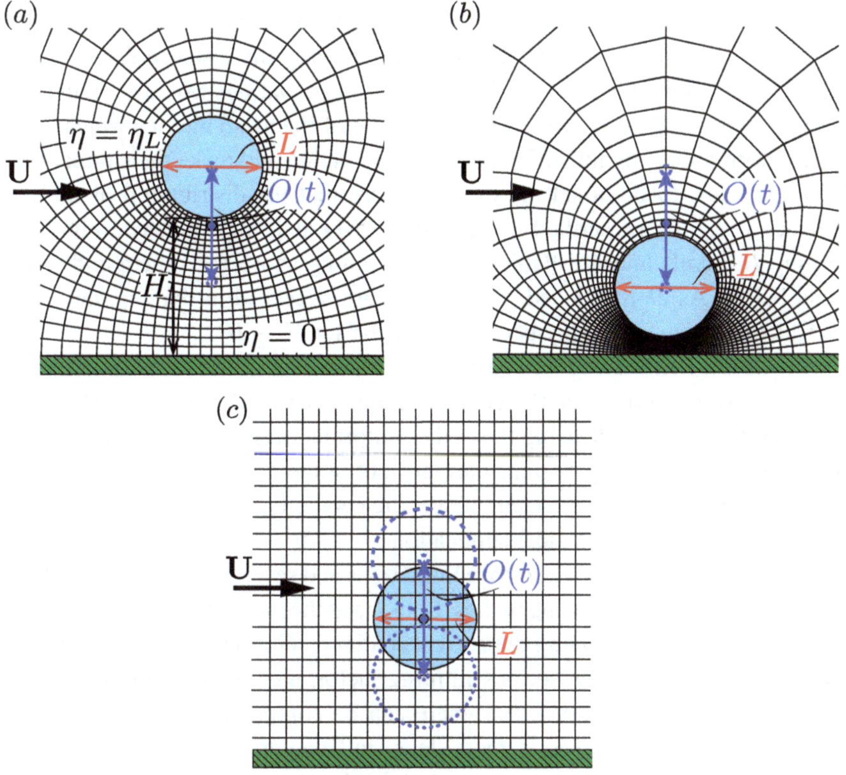

Figure 5.2 Sketch of a sphere of diameter L moving vertically with respect to a horizontal plate at rest: (a) and (b) show the discretization in bispherical coordinates for two different distances of the sphere from the plate. (c) The same setup as (a) but for a regular Cartesian mesh non-body-fitted.

appreciated from Figures 5.1b and 5.2c that the IBM is essentially insensitive to the specific geometrical configuration, while the isogeometric grids of Figures 5.1a and 5.2a,b evidence substantial differences even for a minor change in the setup. It must also be pointed out that the case of Figure 5.2 is exceptional since it is simple enough for an analytic meshing to be possible. In fact, bispherical coordinates can be employed in the form

$$x = a\frac{\sin\xi\cos\phi}{\cosh\eta - \cos\xi}, \qquad y = a\frac{\sin\xi\sin\phi}{\cosh\eta - \cos\xi}, \qquad z = a\frac{\sinh\eta}{\cosh\eta - \cos\xi},$$
$$(5.1)$$

with $0 \leq \xi \leq \pi$, $0 \leq \phi \leq 2\pi$ and $0 \leq \eta \leq \eta_L$. Here the sphere is the coordinate surface at $\eta = \eta_L$ (with center $x = 0$, $y = 0$, $z = a\coth\eta_L$, diameter $L = 2a/\sinh\eta_L$ and distance from the wall $H = a(\cosh\eta_L - 1)/\sinh\eta_L$), while the flat boundary is the isosurface $\eta = 0$.

In fact, for this special coordinate system it would even be possible to prescribe the parameter a as a function of time to vary the distance H between sphere and wall; in this framework, the governing equations would contain additional terms generated by the time derivative of the mesh metrics which, for large deformations, might stiffen the time integration and make the simulation more expensive. Additional disadvantages are that the distribution of nodes in space is rigidly prescribed by equations (5.1) and, for a given resolution, varying H entails uncontrollable mesh coarsening above the sphere and vice versa below. As H is reduced, the isolines are squeezed in the gap, and the coordinate system eventually becomes singular for $H \to 0$.

In contrast, IBMs using a fixed regular mesh avoid all these issues and, in principle, would allow even the contact of the sphere with the plane ($H = 0$); this would be impossible using the body-conformal mesh. The most appealing characteristic is that, by design, they are not subject to the aforementioned limits and are largely independent of specific geometry or flow configurations, making them ideal for practical applications.

On the other hand, IBMs also have limitations and several changes are needed to cope efficiently with moving bodies. Looking at the meshes in Figures 5.1 and 5.2, it can be seen that having the correct spatial resolution at the wall by non-body-fitted grids can be a serious limiting factor, and using a homogeneous mesh with a fine resolution everywhere can be computationally unbearable. This point will be discussed in more detail in Chapter 8. Here we will focus mostly on the technical changes needed by IBMs when the immersed interface moves in time through a fixed mesh and some of the Eulerian nodes change "identity" from solid to fluid and vice versa.

5.1 Cell Tagging with Moving Boundaries

The first obvious change needed by any IBMs dealing with moving boundaries is the node tagging (Chapter 3), which now becomes part of the time integration algorithm. In fact, looking at Figure 5.3 it is clear that the "footprint" of the body on the underlying grid changes in space; therefore Eulerian nodes have a status that depends on time. This entails a substantial computational overhead since, while for fixed boundaries the ray-tracing algorithm of Section 3.2 is only a preprocessing part of the simulation, for moving bodies the ray tracing is performed at every time step and this justifies the need for accelerated tagging algorithms as described in Section 3.3.

It is worthwhile mentioning that the mesh must be tagged at every time step, even if the integration is performed with a Δt small enough to have

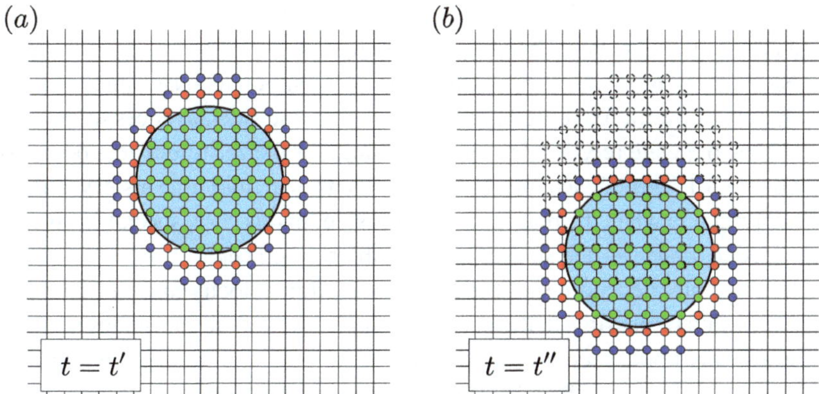

Figure 5.3 (a) Moving object immersed into a regular Cartesian mesh at two different times; the set of internal (green bullets), boundary (red bullets) and external (blue bullets) points changes from time to time. The dashed circles in panel (b) indicate the position of the tagged points at time t'.

displacements of the boundary much smaller than the grid spacing Δ. In fact, even if none of the Eulerian nodes changed status between two consecutive time steps, their relative position from the immersed boundary would change and the same happens to many quantities needed for the time integration. This is the case of the interpolation coefficients for boundary reconstruction in direct methods (Section 4.2.1) as well as the smoothed δ-functions in Lagrangian methods (Section 4.1.1) or the support domain for MLS interpolation (Section 4.2.2).

We can conclude this discussion by stressing that the same tagging procedures used for fixed bodies can be employed for moving boundaries, even if, in the latter case, their computational efficiency becomes a key factor for the feasibility of the simulation.

As mentioned before, another potential issue with moving bodies is the change of status of the grid cells next to the boundary from internal (solid) to external (fluid) and vice versa. Figure 5.4 shows the occurrence of two opposite cases depending on whether the solid boundary advances in the fluid (a) or if the boundary recedes and layers of internal cells become fluid (b).

Usually, the first occurrence is not much of a problem as a fluid point which becomes suddenly internal can be treated in one of the modes detailed in Section 4.3 without affecting the outer flow solution. In contrast, when the boundary recedes there are points whose status changes across two consecutive time steps from internal to interface, or from interface to fluid. These *freshly cleared* cells, depending on the specific IBM, might not have a physical time history to allow the computation of explicit terms of the governing equations on the fluid side and special treatments are needed (Udaykumar et al., 1999).

Note that in most cases, both situations occur simultaneously since, when a body moves over a fixed mesh, as in Figure 5.3, one side advances in the fluid while the opposite recedes. On the other hand, for the zero-thickness bodies of Section 3.3.1, even if each interface point has its own fluid time history, it is not the correct one when the point changes wet side; in this case, special treatments are needed whenever the surface changes the crossed cell.

In principle, depending on the body translation velocity and Eulerian grid size, it would be possible that more than a single layer of internal nodes becomes fluid and this would decrease the accuracy of the simulation since, as we will see, the special treatment of freshly cleared cells essentially consists of extrapolating information from the closest fluid nodes. Considerations on stability of the integration and time accuracy of the solution, however, reveal that the mesh size Δ, maximum velocity of the problem U and time step Δt are not independent and their nondimensional combination $\Delta t U / \Delta$, or the CFL number, must not exceed a stability threshold which is order one. This implies that over the mesh nothing can move more than Δ within a time step (actually, the displacement is much smaller) and the potential problem is therefore practically avoided.

5.2 Diffused- versus Sharp-Interface IBMs

In Chapter 4 we classified IBMs using the broad categories of *continuous* and *direct* approaches, depending on whether the forcing terms were already defined in the initial governing equations or only after some discretization had been performed. In the context of moving boundaries, it is useful to further distinguish the methods between *diffused interface*, when the IB forcing is spread among a few cell layers across the immersed boundary, or *sharp interface* when the boundary reconstruction modifies only a single node next to the interface. In the former methods, boundary discontinuities are enforced only in an integral sense (Udaykumar et al., 2001), such as Peskin's approach of Section 4.1 with the smoothed δ-functions or the direct Lagrangian methods of Section 4.2.2 using either smoothed δ-functions or MLS interpolations. Other Lagrangian methods not discussed in this book are the volume-of-fluid and phase-field (Mittal and Iaccarino, 2005), in which the boundary is identified through a scalar field defined in the whole computational domain and any discontinuities across the interface are smoothed. On the other hand, in those Eulerian methods in which the IB forcing is enforced by modifying the computational stencil near the immersed boundary, such as in Sections 4.2.1, the boundary separates sharply the outer fluid from the inner solid volume and discontinuities in velocity and pressure are maintained across the boundary.

For the first class of methods, the spreading of IB forcing over a few grid cells on both sides of the immersed boundary naturally generates a smooth transition between internal and external nodes, producing the preconditioning needed for grid points transitioning from the inside into the fluid.

The only requirement to be satisfied is that the stencil of the spatial discretizations of the flow solver should be smaller than or equal to the spreading of the interface. In this way, velocity and pressure derivatives can be directly computed in all fluid points close to the moving interface.

According to this discussion, the extension of diffused-interface IBMs to moving bodies is straightforward since no special treatment of the freshly cleared cells is needed. The only change needed is the update of the grid tagging and the recomputation of the coefficients for the boundary reconstruction at every time step.

Conversely, for sharp-interface methods those grid points that have just emerged into the flow from the body inside do not have a prior history: For these freshly cleared cells the time advancement scheme of the fluid solver (which typically involves the flow solution at the previous time steps) can not be computed in a straightforward way and ad hoc treatment is needed.

5.3 Field-Extension for Sharp-Interface IBMs

Here we consider the specific case of a sharp-interface IBM in which the forcing is applied at the first external node (see Section 4.2.1) and the fluid solver is based on central second-order finite differences (as the code made available with the present book). The time advancement of the Navier–Stokes equations from the time level t^n to the next t^{n+1}, requires velocity and pressure, as well as their derivatives in the momentum equation, to be evaluated at time $t \leq t^n$ (see equation (4.41)), to compute all the explicit terms of the scheme. However, as the boundary moves through the fixed grid, some Eulerian points next to the interface change status across two consecutive time steps, and their variables at t^n are not physically consistent with the new domain at t^{n+1}. Figure 5.4 shows the opposite cases of a boundary advancing (panels (a)) and receding (panels (b)) from the fluid phase. In the first case (Figure 5.4a), forcing points at t^n become inner points at t^{n+1} (filled triangles) and the neighboring fluid points (empty triangles) become forcing points; in both cases no special treatment is needed as, even if the new inner points (filled triangles) at t^{n+1} do not possess the correct physics, they do not contaminate the outer flow solution. On the

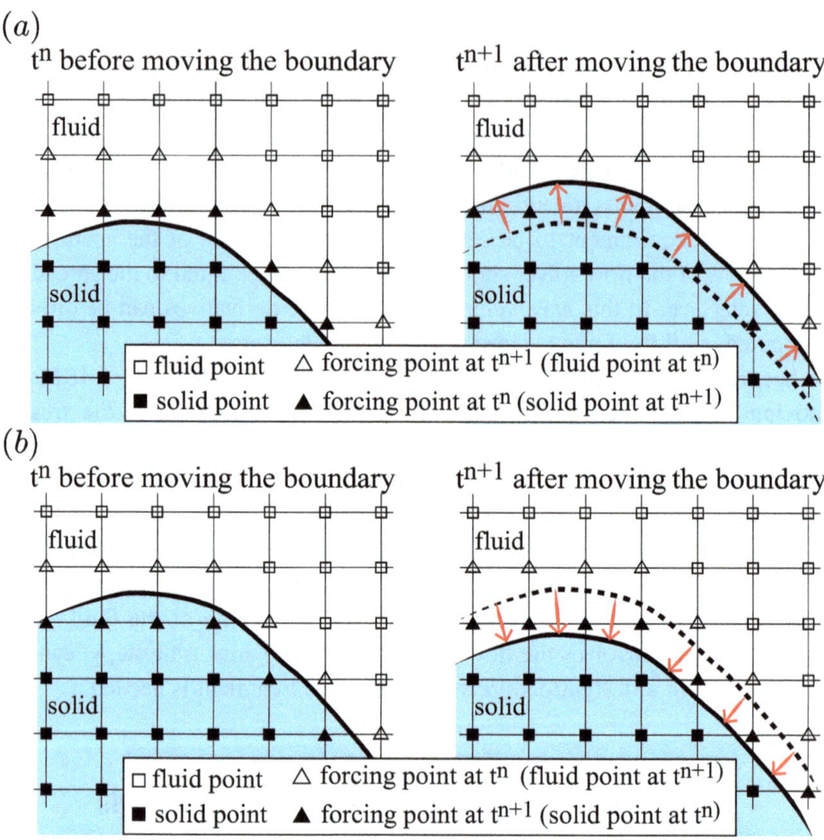

Figure 5.4 (a) Solid phase advances over the fluid phase. (b) Solid phase withdraws from the fluid phase.

other hand, the new forced points (open triangles) remain in the fluid phase and anyway attain variable values at t^{n+1} given by boundary reconstruction or IB forcing scheme rather than by the local Navier–Stokes solution.

In contrast, when the boundary recedes from the fluid (Figure 5.4b) solid points become forcing nodes (solid triangles) and these in turn become standard fluid points (open triangles). The former transition is similar to the previous case and does not imply any numerical issue, as the IB forcing overwrites any inconsistency in the evaluation of the flow derivatives. Unfortunately, the solution of equations at the newly emerged fluid points (open triangles) involves velocity derivatives from previous times that are incorrect for all $t \leq t^n$ when the point was not evolved through the equations. Indeed, it should be noted that forcing points do have the "correct" values for the velocity and pressure, since

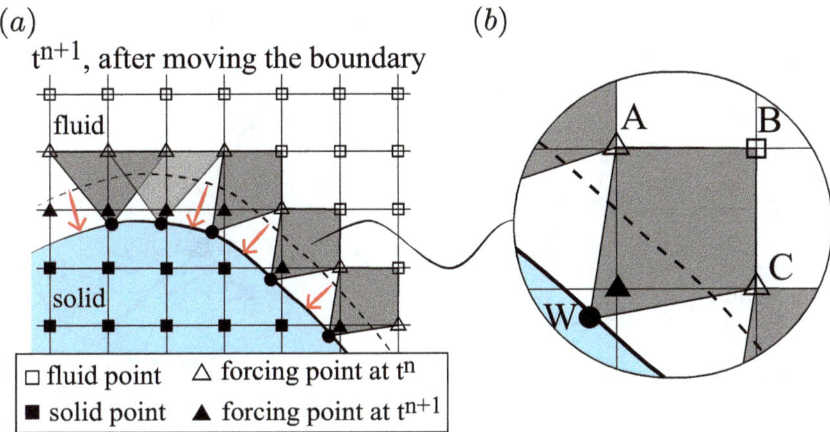

Figure 5.5 (a) Creation of *freshly cleared* cells on a fixed Cartesian grid due to boundary motion and corresponding stencil to interpolate the flow variables from neighboring nodes, which is better visualized in the inset (b).

they come from IB forcing and boundary reconstruction and thus they satisfy the boundary condition at the immersed interface. Yet, the same nodes have incorrect spatial derivatives of those variables since internal nodes from the solid phase have been involved, as for the forcing points at t^{n+1} (filled triangles). These unphysical values can introduce spurious gradients and vorticity at the boundary, yielding large errors and destabilizing the solution.

A possible trick to avoid this issue consists of estimating the missing information through interpolations or extrapolations from the appropriate neighboring points just before it is needed (see Mittal & Iaccarino (2005)). For example, let us consider the points marked by open triangles in Figure 5.4b, which are forced nodes at t^n but become external at t^{n+1} with a solution which is to be determined by the Navier–Stokes equations. The explicit terms therein entail velocity derivatives at t^n, in turn involving velocities at the forcing points (filled triangles), which at t^n were solid points.

As a freshly cleared grid point emerges, its velocity component $u_{\blacktriangle}$, at time t^n, can be reconstructed from the neighboring fluid cells, as shown in Figure 5.5, where the grey area indicates the stencil of the bilinear interpolant:

$$u_{\blacktriangle} = \phi_A u_A + \phi_B u_B + \phi_C u_C + \phi_W u_W. \tag{5.2}$$

The interpolation coefficients are obtained by imposing that the value of the interpolant in A, B, C, W is equal to u_A, u_B, u_C, u_W. If the discretization is performed on a colocated grid, then the same coefficients can be retained for all velocity components and pressure. In contrast, if a staggered mesh is

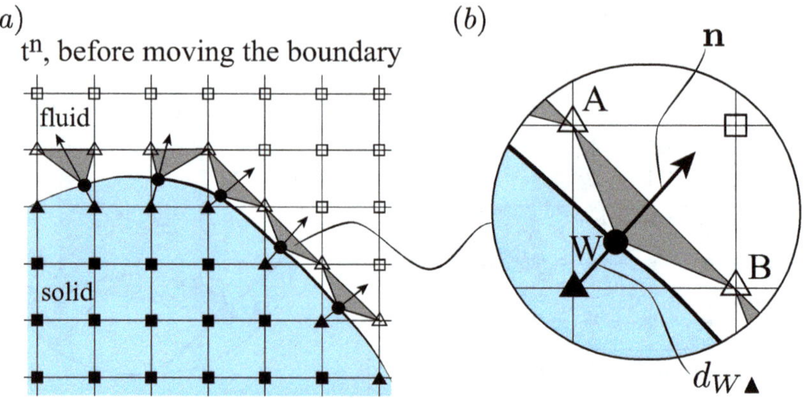

Figure 5.6 (a) Sketch of field extension into the first internal node layer and stencil needed for the field extrapolation. (b) Magnification of the stencil to extrapolate flow variables from neighboring fluid nodes into the body.

employed, each quantity needs its interpolation coefficients to account for the different relative positions of boundary and discretized unknowns.

With the solution at time t^n extended with the variable from equation (5.2), it is now possible to advance the equations at the time level t^{n+1} although only with a first-order accuracy in time, which in many cases is not enough. Unfortunately, to obtain higher time accuracies, such as in the second-order explicit Adams–Bashforth scheme or the progeny of the multistep methods, information for the freshly cleared nodes is needed at times $t < t^n$, when they are still inside the body.

A possible remedy for this issue, widely used for the second-order Adams–Bashforth scheme, is the *field-extension* strategy proposed by Yang and Balaras (2006). The main idea is to *extend* velocity and pressure into the first node layer of the immersed body at the end of each time step, regardless of the fate of such points. In particular, the variables are extrapolated at all Eulerian internal nodes that have at least one neighbor in the fluid phase (filled triangles in Figure 5.6). Indeed, if the boundary withdraws from the fluid, these solid points will turn into forcing points and belong to the stencil for evaluation of flow derivatives at the actual forcing points (empty triangles). This is achieved, at any t^n time step, by a bilinear extrapolation involving the neighboring fluid points and a Lagrangian boundary node along the normal direction as shown in Figure 5.6:

$$u_{\blacktriangle} = \phi_A u_A + \phi_B u_B + \phi_W u_W,\tag{5.3}$$

where, as for equation (5.2), the extrapolation weights are obtained requiring that the value of the bilinear approximation in A, B, W matches u_A, u_B, u_W, respectively (see equation (4.18)). In this way, not only are the flow derivatives at the forcing points (empty triangles) at the time level t^n physically meaningful, but also those at the previous time levels. This cost-efficient strategy is very robust (Yang and Balaras, 2006) and it allows the use of second-order multistep formulas such as Adams–Bashforth.

It is worthwhile noticing that, for a specific combination of wall reconstruction and boundary condition, the extension of the pressure field can be achieved directly from the governing equations, without resorting to any extrapolation. In fact, in Figure 5.6 let $d_{W\blacktriangle}$ be the distance between a solid grid point closest to the boundary (filled triangles) and the corresponding Lagrangian node W in the boundary normal direction $\mathbf{n}$. The momentum equation along $\mathbf{n}$ reads

$$\frac{\mathrm{D}\mathbf{u}}{\mathrm{D}t} \cdot \mathbf{n} = -\nabla p \cdot \mathbf{n} + \frac{1}{Re}\nabla^2 \mathbf{u} \cdot \mathbf{n}. \tag{5.4}$$

Observing that viscous terms vanish for IBMs based on a linear interpolation rule and that the fluid velocity at the boundary is equal to that of the body surface (no-slip condition), equation (5.4) reduces to

$$\frac{D^2\mathbf{X}_W}{Dt^2} \cdot \mathbf{n} = -\left.\frac{\partial p}{\partial n}\right|_W, \tag{5.5}$$

where the left-hand side is the local surface acceleration along $\mathbf{n}$ and the right-hand side is the pressure derivative in the same direction. Hence, the extended pressure is readily obtained from a finite-difference approximation for the pressure term:

$$\frac{p_W - p_{\blacktriangle}}{d_{W\blacktriangle}} = -\frac{D^2\mathbf{X}_W}{Dt^2} \cdot \mathbf{n} \quad \Rightarrow \quad p_{\blacktriangle} = p_W + d_{W\blacktriangle}\frac{D^2\mathbf{X}_W}{Dt^2} \cdot \mathbf{n}. \tag{5.6}$$

In this discussion, we have considered only the case of centered second-order spatial schemes, in which only the first neighboring points are needed for the flow derivatives. When higher-order schemes are adopted, the larger stencil amplifies the issues of freshly cleared nodes and the derivatives at the boundary involve several layers within the solid. The higher the accuracy, the more solid points are involved in the discrete derivative. In principle, this could be solved by *propagating* further the field extension into the solid to determine the flow velocity over all the needed points. However, to match the accuracy of the flow solver, this would entail high-order polynomials extrapolating at more distant locations, which is prone to yield large numerical errors and instabilities.

As a matter of fact, despite many attempts to develop high-order IBMs, computations for complex geometric configurations with moving boundaries are usually produced with at most second-order flow solvers.

6

Hydrodynamic Loads Computation

An important aspect of computational fluid dynamics is the evaluation of the forces exchanged between boundary and flow; this is key information during design, for example to minimize drag or maximize the load along a specific direction, or in applications involving fluid–structure interaction (FSI), where body dynamics are determined by the flow and vice versa (Sections 7.2 and 7.3). In the context of IBMs, as the boundary does not coincide with coordinate surfaces of the mesh, the evaluation of hydrodynamic loads can be performed in several ways, each one entailing different computational costs and degrees of precision.

The local traction (force per unit area) $\mathbf{t}_f$ on a wet boundary with an outward normal $\mathbf{n}$ is given by the sum of pressure and viscous contributions according to

$$\mathbf{t}_f = \mathbf{T} \cdot \mathbf{n} = -p\mathbf{n} + \tau \cdot \mathbf{n}, \tag{6.1}$$

where $\mathbf{T}$ is the total hydrodynamic stress tensor, p is the pressure and τ is the viscous stress tensor. The latter, for an incompressible flow and a Newtonian fluid with dynamic viscosity μ (see equation (2.18)), reads $\tau = \mu(\nabla\mathbf{u} + (\nabla\mathbf{u})^T)$.

In the case of zero-thickness bodies, surfaces are open and wet on both sides; therefore the hydrodynamic loads are produced by the positive $\mathbf{n}^+$ and negative $\mathbf{n}^- = -\mathbf{n}^+$ sides of the boundary:

$$\mathbf{t}_f = [-(p^+ - p^-)\mathbf{n}^+ + (\tau^+ - \tau^-) \cdot \mathbf{n}^+], \tag{6.2}$$

where p^+ (p^-) and τ^+ (τ^-) are the corresponding pressure and viscous stress tensor on the positive (negative) wet side. Given the local traction $\mathbf{t}_f$, the corresponding hydrodynamic force acting on a surface dA is thus $d\mathbf{F} = \mathbf{t}_f dA$.

98

6.1 Action and Reaction

A theoretical bonus of IBMs is that hydrodynamic loads can be obtained directly as a part of the solution without any additional computations. In fact, the essence of these methods is to replace the presence of the boundary with the forces they exert on the flow. We indicate these by $\mathbf{f}_{IB}$ as the IB contribution to the total volume forces $\mathbf{f}$ of equation (4.1); according to Newton's third law of motion, the equal and opposite reaction is the force (per unit mass) of the flow on the boundary $-\mathbf{f}_{IB}$ which is readily available for all points of body surface.

This method, sketched in Figure 6.1, applies to all IBMs and, upon summation of the contributions from all forced Eulerian cells, yields the resultant of the total force and its moment with respect to any points if needed.

In principle, this procedure is effective and inexpensive as it does not add any computational overheads; however, there are also disadvantages and in practice different procedures are more often employed.

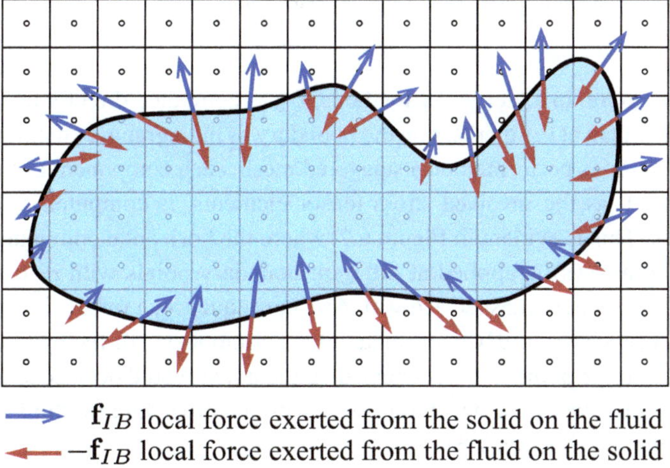

Figure 6.1 Sketch of the forces exchanged between the solid and the fluid: blue open arrows (solid to fluid) and red filled arrows (fluid to solid).

One problem is that $\mathbf{f}_{IB}$ accounts for the *total* force without separating pressure and viscous components: Distinguishing between the two contributions, however, is relevant both for a deeper understanding of the flow physics as well as for applications related to flow control and shape optimization for drag reduction. Another issue is that with this procedure it is not possible to separate the loads applied on positive and negative sides (with respect to the normal direction) of the boundary.

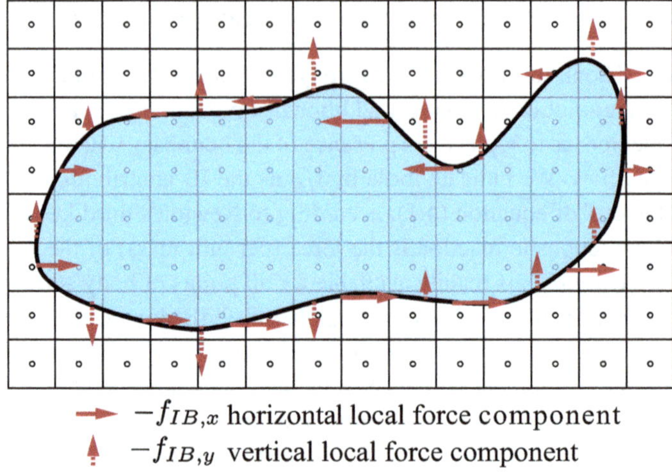

Figure 6.2 Sketch of the application points of the horizontal and vertical components of the force exerted from the fluid on the solid.

Furthermore, in Eulerian IB methods, $\mathbf{f}_{IB}$ is computed directly on the fixed grid rather than at the markers on the boundary as in Lagrangian methods; if the discretization is performed on a staggered mesh, each component of the vector $\mathbf{f}_{IB}$, as well as the involved stress tensor elements, is computed at different positions. This is evident in Figure 6.2, where the horizontal component of the $\mathbf{f}_{IB}$ (solid arrows) is applied at different boundary points with respect to the vertical component (dashed arrows). As a consequence, a further interpolation step is necessary in order to evaluate all the components of the hydrodynamic loads at a single point on the immersed boundary, as would be needed to couple the flow solution with a structural solver within a procedure for fluid–structure interaction (see Section 7.3).

6.2 Bilinear Interpolation

According to the previous arguments, hydrodynamic forces are often computed using a reconstruction procedure, similar to that of Chapter 4, applied directly to the primitive variables. Figure 6.3 shows, for a two-dimensional example, a bilinear interpolation for a staggered discretization. Black bullets on the immersed boundary indicate the Lagrangian markers where forces have to be evaluated, and for each one the corresponding stencil of the bilinear interpolation is reported. The inset of Figure 6.3 shows the interpolation scheme for the

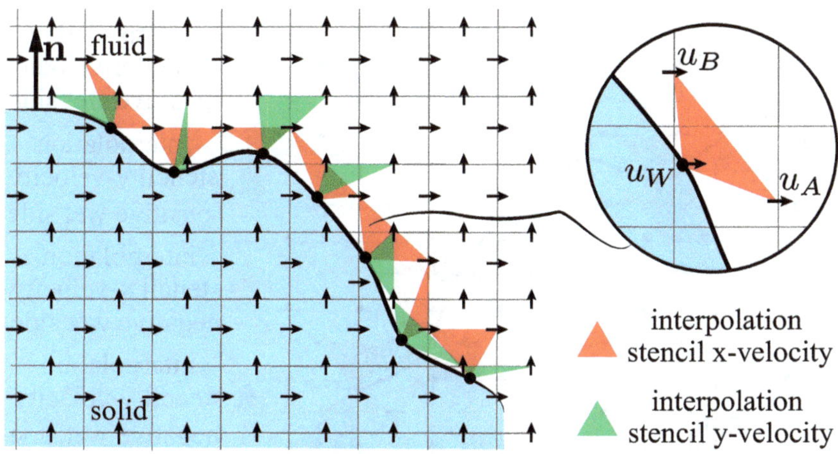

Figure 6.3 Bilinear interpolation procedure to calculate the hydrodynamic stress acting on the immersed body.

horizontal velocity component in the vicinity of the Lagrangian marker, which can be written as

$$u = \zeta_1 x + \zeta_2 y + \zeta_3, \tag{6.3}$$

with the interpolation weights such that

$$\begin{aligned} u_A &= \zeta_1 x_A + \zeta_2 y_A + \zeta_3, \\ u_B &= \zeta_1 x_B + \zeta_2 y_B + \zeta_3, \\ u_W &= \zeta_1 x_W + \zeta_2 y_W + \zeta_3. \end{aligned} \tag{6.4}$$

Here, velocities u_A and u_B come directly from the flow solution, while u_W is known as the boundary condition or is determined through additional models, such as the structural dynamics in FSI applications (see Section 7.3).

Differentiating equation (6.3) with respect to the spatial coordinates yields

$$\frac{\partial u}{\partial x} = \zeta_1 \text{ and } \frac{\partial u}{\partial y} = \zeta_2. \tag{6.5}$$

A similar reconstruction applied to the vertical velocity component (green interpolation stencil in Figure 6.3) provides the partial derivatives of the v component, and therefore the remaining terms of the velocity gradient tensor from which viscous stresses are obtained.

It should be noted that the shape and size of the interpolation stencil depend on the marker location with respect to the fluid cells. This observation clarifies that, even if all interpolants are second-order accurate, the precision of the interpolation varies among the various Lagrangian markers, as it depends on the distance of the Lagrangian marker from the surrounding Eulerian grid points.

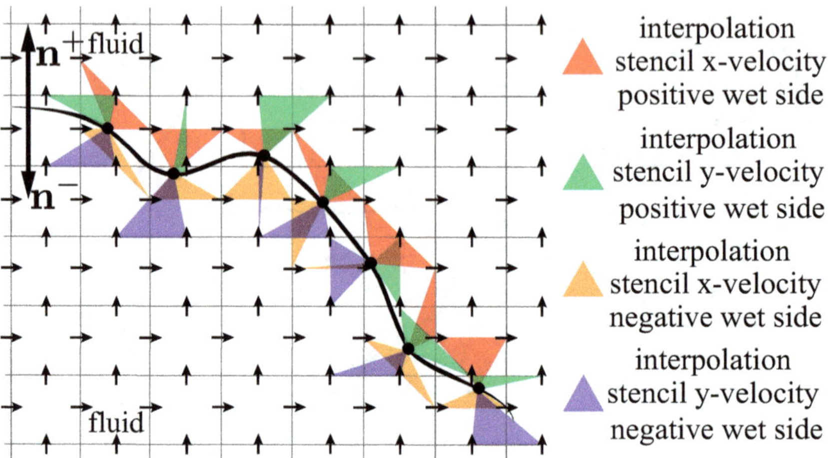

Figure 6.4 Bilinear interpolation procedure for a thin body that is wet on both sides.

Furthermore, as each Eulerian quantity "lives" on its own staggered grid, the relative position of any of the Lagrangian markers with respect to the Eulerian mesh is different, implying that also the precision of the interpolation changes for the three components of the vector $d\mathbf{F}$ evaluated on the same marker. This is evident in Figure 6.3, which highlights the different interpolation stencils for horizontal (red triangles) and vertical (green triangles) components.

In the case of zero-thickness IB, such as a membrane, the hydrodynamic force has to be computed along both positive $\mathbf{n}^+$ and negative $\mathbf{n}^- = -\mathbf{n}^+$ normal directions (see equation (6.2)). As shown in Figure 6.4, for a given Lagrangian marker, the derivatives of the velocity components are computed by using interpolation stencils on positive (red and green triangles) and negative (orange and purple triangles) wet sides. Since the marker is generally not equidistant from the velocity (as well as pressure) locations on the two sides of the IB, the resulting interpolation stencils are not symmetric with respect to the wet surface; this implies a different precision of the gradient evaluation between the positive and negative sides.

A similar interpolation procedure could also be applied to evaluate the pressure on the wet surface. Unlike the velocity field, however, pressure is unknown at the boundary since, for incompressible flows, it does not come as a boundary condition or as part of the solution of the structural solver in FSI. Hence, the bilinear interpolant should be based on three pressure values obtained from the Navier–Stokes solution. Referring to Figure 6.5a:

$$p_W = \zeta_1 x_W + \zeta_2 y_W + \zeta_3, \tag{6.6}$$

where the coefficients $\zeta_{1,2,3}$ are obtained by imposing $p_A = \zeta_1 x_A + \zeta_2 y_A + \zeta_3$ and the same conditions on B and C. The evaluation of the boundary pressure through (6.6) thus corresponds to an extrapolation, which can result in large errors.

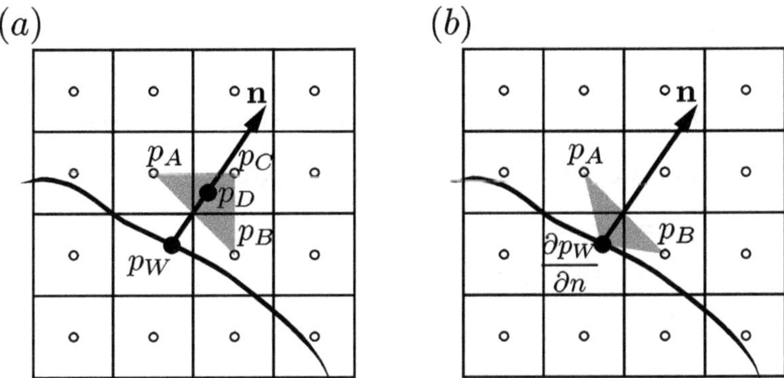

Figure 6.5 Interpolation stencil for the bilinear interpolation of pressure using (a) the point D and (b) the Neumann condition for pressure at the wall.

This issue can be circumvented by evaluating the pressure at point D (still indicated in Figure 6.5a), which is selected closer than a local mesh size from W along the normal direction and within the interpolation stencil of equation (6.6). Once p_D is computed by interpolation, p_W is obtained by solving the Navier–Stokes equation, in the boundary-layer approximation, projected in the wall-normal direction (see equation (5.6)):

$$p_W = p_D + d_{DW} \frac{D^2 \mathbf{X}_W}{Dt^2} \cdot \mathbf{n}, \tag{6.7}$$

where $\mathbf{X}_W$ is the instantaneous position of the W Lagrangian marker and d_{DW} is the distance between D and W locations.

Alternatively, the bilinear interpolation (6.6) can be built directly using equation (6.7), by changing the interpolation stencil as in Figure 6.5b (Yang and Balaras, 2006).

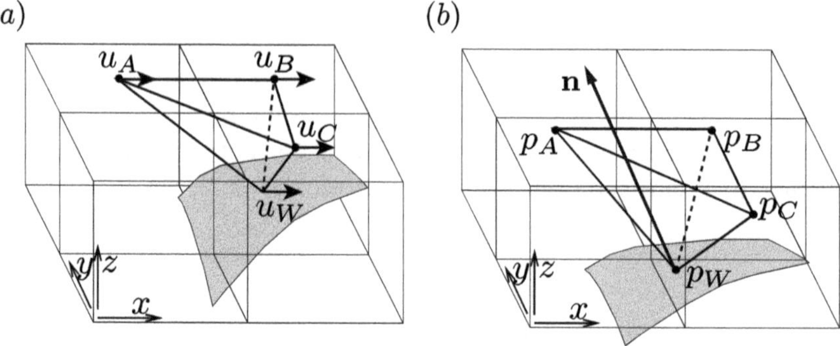

Figure 6.6 Trilinear interpolation scheme for (a) velocity and (b) pressure.

The interpolation weights are thus obtained by imposing

$$p_A = \zeta_1 x_A + \zeta_2 y_A + \zeta_3,$$

$$p_B = \zeta_1 x_B + \zeta_2 y_B + \zeta_3,$$

$$\left.\frac{\partial p}{\partial n}\right|_W = \underbrace{\left.\frac{\partial p}{\partial x}\right|_W}_{\zeta_1} n_x + \underbrace{\left.\frac{\partial p}{\partial y}\right|_W}_{\zeta_2} n_y = -\frac{D^2 \mathbf{X}_W}{Dt^2} \cdot \mathbf{n}, \tag{6.8}$$

and the pressure on the boundary is obtained by evaluating the resulting bilinear interpolant in W.

The same procedures can be extended to the three-dimensional case in a straightforward manner. As shown in Figure 6.6a, a trilinear interpolation is used to evaluate the velocity gradient:

$$u = \zeta_1 x + \zeta_2 y + \zeta_3 z + \zeta_4, \tag{6.9}$$

where the unknown coefficients are determined by imposing that the interpolant is equal to u_A, u_B, u_C and u_W in A, B, C and W, respectively. Differentiating equation (6.9) with respect to the spatial directions it results: $\frac{\partial u}{\partial x} = \zeta_1$, $\frac{\partial u}{\partial y} = \zeta_2$ and $\frac{\partial u}{\partial z} = \zeta_3$. On the other hand, Figure 6.6b shows interpolation stencil for the boundary pressure with the coefficients determined by imposing

$$p_A = \zeta_1 x_A + \zeta_2 y_A + \zeta_3 z_A + \zeta_4,$$

$$p_B = \zeta_1 x_B + \zeta_2 y_B + \zeta_3 z_B + \zeta_4,$$

$$p_C = \zeta_1 x_C + \zeta_2 y_C + \zeta_3 z_C + \zeta_4,$$

$$\left.\frac{\partial p}{\partial n}\right|_W = \underbrace{\left.\frac{\partial p}{\partial x}\right|_W}_{\zeta_1} n_x + \underbrace{\left.\frac{\partial p}{\partial y}\right|_W}_{\zeta_2} n_y + \underbrace{\left.\frac{\partial p}{\partial z}\right|_W}_{\zeta_3} n_z = -\frac{D^2 \mathbf{X}_W}{Dt^2} \cdot \mathbf{n}. \tag{6.10}$$

6.3 Smooth Hydrodynamic Loads

The previous procedures are quite robust and relatively inexpensive, although they are affected by the annoying issue that the local loads can oscillate from marker to marker as the distance from the closest Eulerian fluid node varies randomly along the immersed boundary, and so does the amplitude of the numerical truncation error of the various terms. This might not be much of a problem for massive rigid objects whose dynamics depend only on the *resultant* loads and the high-frequency temporal oscillations are filtered out by the body inertia. In contrast, for low-mass, deformable bodies, the irregular distribution of the surface loads turns out to be catastrophic as *local* distributions of forces and moments determine the instantaneous body configuration, in turn affecting the flow through fluid–structure interaction mechanisms: This coupling usually leads to the divergence of the integration.

In order to avoid this issue, Vanella and Balaras (2009) proposed the use of the same MLS interpolation employed for the IB forcing that was introduced in Section 4.2.2 to compute the hydrodynamic loads at the Lagrangian markers. For this purpose, however, the support domain centered at $\mathbf{X}_l$ (Figure 4.9b), used to compute the IB forcing, is not appropriate as it encloses nodes on both sides of the boundary. In fact, if the immersed body has a solid interior, the resulting hydrodynamic loads would be contaminated by the unphysical internal nodes while, for zero-thickness bodies like a membrane, stresses on the two wet surfaces are generally different and they must be computed separately, using only quantities from the pertaining side.

This problem can be addressed by casting, from each force node, a segment of length h (proportional to the grid size) along the local boundary normal (Figure 6.7a); the distal edge of this segment defines the "probe" position $\mathbf{X}_p$ where the needed quantities, pressure and velocity gradient components, are reconstructed via MLS interpolation:

$$p(\mathbf{X}_p) = \sum_{k=1}^{N_k} \phi_k(\mathbf{X}_p) p(\mathbf{x}_k),$$

$$\left. \frac{\partial u_i}{\partial x_j} \right|_{\mathbf{x}=\mathbf{X}_p} = \sum_{k=1}^{N_k} \left. \frac{\partial \phi_k}{\partial x_j} \right|_{\mathbf{x}=\mathbf{X}_p} u_i(\mathbf{x}_k),$$

(6.11)

where ϕ_k are the MLS interpolation weights, following the same notation as in Section 4.2.2. It has to be remarked that since the interpolation always involves a cloud of N_k neighbouring points and the derivatives $\partial \phi_k / \partial x_j$ are computed analytically, the velocity gradient components are smooth at the boundary.

Nevertheless, pressure and velocity gradients from equation (6.11) are evaluated at the probe location rather than on the immersed body. As already done in equations (5.6) and (6.7), the pressure at the Lagrangian marker can be obtained from the fluid pressure through a boundary-layer approximation:

$$p|_W = p(\mathbf{X}_l) = p(\mathbf{X}_p) + h\frac{D\mathbf{U}_W}{Dt} \cdot \mathbf{n} + O(h^2), \tag{6.12}$$

where $\mathbf{U}_W$ and $\mathbf{n}$ are the local boundary velocity and outward normal.

The velocity gradient at the body surface is instead assumed equal to that at the probe (equation (6.11)), which is a good approximation provided the grid is sufficiently refined in the near-wall region (de Tullio and Pascazio, 2016):

$$\left.\frac{\partial u_i}{\partial x_j}\right|_W = \left.\frac{\partial u_i}{\partial x_j}\right|_P + O(h) = \sum_{k=1}^{N_e} \left.\frac{\partial \phi_k}{\partial x_j}\right|_P u_i(\mathbf{x}_k) + O(h), \tag{6.13}$$

where the subscripts W and P indicate the wall ($\mathbf{X}_l$) and probe ($\mathbf{X}_p$) positions. The second-order accuracy of the discretization scheme is preserved provided the fluid velocity next to the surface varies linearly (which is certainly true if the first fluid node is within the viscous sublayer).

Combining pressure (6.12) and velocity gradient (6.22) contributions, the surface hydrodynamic stress tensor is evaluated as

$$\tau|_W = -p|_W\mathbf{I} + \mu\left(\nabla\mathbf{u} + (\nabla\mathbf{u})^T\right)\Big|_W \approx -\left(p|_P\mathbf{I} + h\frac{D\mathbf{U}_W}{Dt} \cdot \mathbf{n}\right) + \mu\left(\nabla\mathbf{u} + (\nabla\mathbf{u})^T\right)\Big|_P. \tag{6.14}$$

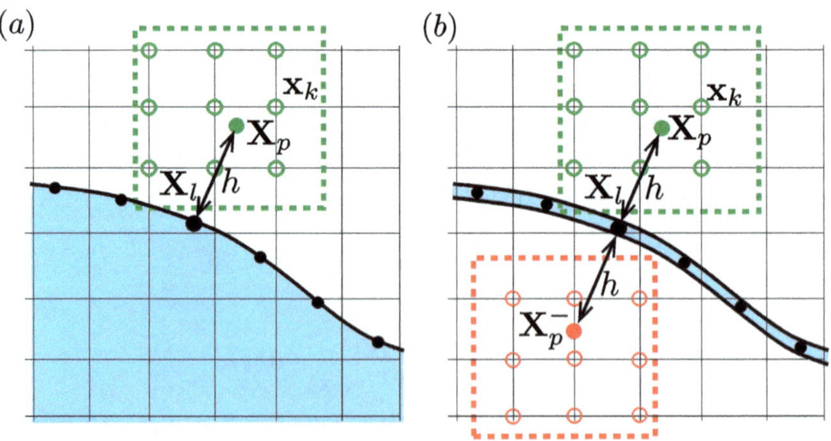

Figure 6.7 Sketch of the support domains and probes for the evaluation of the hydrodynamic loads according to the MLS method: (a) standard body with an inner solid region; (b) zero-thickness immersed boundary.

A higher-order estimate of the hydrodynamic loads can be obtained by expanding pressure in Taylor series:

$$p|_W = p|_P - \left.\frac{\partial p}{\partial n}\right|_W h - \frac{1}{2}\left.\frac{\partial^2 p}{\partial n^2}\right|_W h^2 + O(h^3), \tag{6.15}$$

where the first pressure derivative is obtained by the boundary acceleration (see equation (5.5)), and the second one can be estimated through the following finite difference:

$$\left.\frac{\partial^2 p}{\partial n^2}\right|_W = \left(\left.\frac{\partial p}{\partial n}\right|_P - \left.\frac{\partial p}{\partial n}\right|_W\right)\frac{1}{h} + O(h), \tag{6.16}$$

which injected into equation (6.15) yields

$$p|_W = p|_P - \frac{1}{2}\left(\left.\frac{\partial p}{\partial n}\right|_W + \left.\frac{\partial p}{\partial n}\right|_P\right)h + O(h^3). \tag{6.17}$$

The term $\left.\frac{\partial p}{\partial n}\right|_P$ is thus the pressure derivative in the wall-normal direction evaluated at the probe location, which can be computed as $\nabla p|_P \cdot \mathbf{n}$, where the components of the pressure gradient are obtained by taking the derivatives of the pressure MLS interpolant, similarly to the calculation of the velocity derivatives in equation (6.11). Equation (6.17) was determined by Wang et al. (2019) by assuming a quadratic behavior for the pressure between the wall and probe location. In the same way, they obtained a higher-order expression for the viscous stress valid for rigid immersed bodies in the boundary layer approximation, by using a cubic polynomial to reconstruct the tangential velocity u_ξ:

$$u_\xi(\eta) = d + c\eta + \frac{b}{2}\eta^2 + \frac{a}{6}\eta^3, \tag{6.18}$$

where η is the wall-normal direction. The unknown coefficients a, b, c and d are determined by imposing the no-slip condition on $u_\xi(\eta)|_W$ at the body surface and the local flow information at the probe $\partial u_\xi/\partial\eta|_P$, $\partial^2 u_\xi/\partial\eta^2|_P$ and $\partial^3 u_\xi/\partial\eta^3|_P$. These quantities can be obtained from the numerical flow solution (and its derivatives) interpolated in P. The corresponding viscous stress at the wall (in the direction of the tangential velocity u_ξ) is promptly obtained by taking the first derivative of equation (6.18) at the wall:

$$\tau_W = \mu\left.\frac{\partial u_\xi}{\partial\eta}\right|_W = \mu\left.\frac{\partial u_\xi}{\partial\eta}\right|_P - \mu\left.\frac{\partial^2 u_\xi}{\partial\eta^2}\right|_P h + \frac{\mu}{2}\left.\frac{\partial^3 u_\xi}{\partial\eta^3}\right|_P h^2 + O(h^3). \tag{6.19}$$

Alternatively, a higher-order expression for both pressure and the viscous stress tensor, also valid for deformable bodies, can be obtained by writing a truncated Taylor series in the wall-normal direction; this time from the probe to the wall. For instance, the viscous stress components can be Taylor expanded as:

$$\left.\frac{\partial u_i}{\partial x_j}\right|_W = \left.\frac{\partial u_i}{\partial x_j}\right|_P - \left.\frac{\partial(\partial u_i/\partial x_j)}{\partial n}\right|_P h + \frac{1}{2}\left.\frac{\partial^2(\partial u_i/\partial x_j)}{\partial n^2}\right|_P h^2 + O(h^3),$$

$$(6.20)$$

where the normal derivatives appearing in the first- and second-order Taylor corrections are obtained from the flow solution at the probe location. As an example, the first-order term can be written using the gradient of the viscous stress components $\nabla(\partial u_i/\partial x_j)|_P \cdot \mathbf{n}$. The components of this tensor (involving the second derivatives of the velocity field) can be obtained either by MLS-interpolating the second derivatives evaluated in the Eulerian support points:

$$\left.\frac{\partial^2 u_i}{\partial x_j^2}\right|_P = \sum_{k=1}^{N_k} \phi_k(\mathbf{X}_p)\left.\frac{\partial^2 u_i}{\partial x_j^2}\right|_{\mathbf{x}_k}, \qquad (6.21)$$

or by taking the derivative of the MLS-function, interpolating the first derivatives at the Eulerian points:

$$\left.\frac{\partial^2 u_i}{\partial x_j^2}\right|_P = \sum_{k=1}^{N_k} \left.\frac{\partial \phi_k}{\partial x_j}\right|_P \left.\frac{\partial u_i}{\partial x_j}\right|_{\mathbf{x}_k}, \qquad (6.22)$$

similar to what has been done in equation (6.11) for obtaining the first derivatives of the velocity field. This procedure thus allows reconstructing the whole stress tensor at the wall, matching the same accuracy of the underlying fluid solver.

If flow stresses need to be computed on both sides of the wet surface (as for a flapping flag or for the motion of cardiac valve leaflets), the same procedure is applied to a probe placed along the negative normal direction (Figure 6.7b) and each term will have contributions from both sides.

6.4 Validation: Prescribed Kinematics

Before concluding this chapter, we briefly show a validation of the procedure for the load computation for a flow with a body moving with a prescribed kinematics. For this class of problems, a popular test case is the flow around a two-dimensional circular cylinder in a uniform horizontal flow harmonically oscillating in the cross-stream (vertical) direction. It is worth mentioning that this problem could also be solved on a fixed Eulerian grid using a noninertial reference frame moving with the cylinder: In this case, the cylinder at rest would be subjected to an oscillating flow with noninertial forces.

If D is the cylinder diameter, U is the unperturbed free-stream velocity and $y(t) = A\sin(2\pi f t)$ is the oscillation law, then the governing parameters are the Reynolds number $Re = UD/\nu$, the ratio A/D and the Keulegan–Carpenter number $KC = U/(fD)$ that in the present case have been set to $Re = 185$, $A/D = 0.2$ and $KC = 6.4$ corresponding to $f/f_0 = 0.8$ with f_0 the natural shedding frequency in the absence of oscillations.

In this case, the kinematics of the body is prescribed and the no-slip condition should be imposed on the wet boundary of the moving body:

$$\mathbf{u}(\mathbf{X}(t)) = \left.\frac{\mathrm{d}\mathbf{X}(t)}{\mathrm{d}t}\right|_{\mathbf{X}(t)} \quad \text{with } \mathbf{X}(t) \in \partial\Omega(t), \tag{6.23}$$

where $\mathbf{u}$ is the fluid velocity and $\mathbf{X}(t)$ is the instantaneous position of a point belonging to the wet boundary $\partial\Omega(t)$.

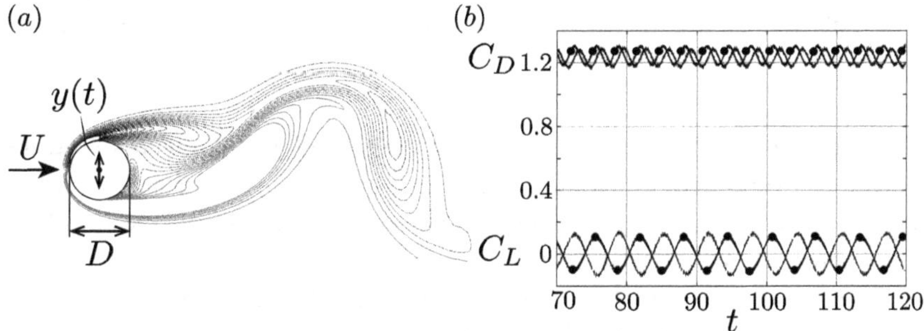

Figure 6.8 (a) Schematic of the flow and snapshot of the spanwise vorticity at the lowest position of the oscillation: ——— for positive, – – – – for negative values. $\Delta\omega = \pm 0.5$. (b) Time evolution of drag (C_D) and lift (C_L) coefficients: ——— present results, ——— with • simulation by Yang and Balaras (2006).

The simulation is performed in a two-dimensional domain of $32D \times 50D$ in the spanwise and streamwise directions with a grid of 300×420 nodes. The mesh distribution is non-uniform and such that in the region around the cylinder the mesh spacing is constant and approximately equal to $0.01D$. An instantaneous snapshot of the spanwise vorticity field is shown in Figure 6.8a together with a schematic of the problem.

Quantitative information is given in Figure 6.8b showing the time history of lift and drag coefficients (computed by using the MLS procedure of Section 6.3) compared again with the results of a similar simulation performed by Yang and Balaras (2006). Both lift and drag coefficients show excellent agreement in magnitude and time periodicity, while the phase shift is simply due to a random initial onset of the von Kármán instability caused by the perturbations of the different numerical methods.

Figure 6.9 shows the local wall distributions of pressure coefficient and vorticity during the instant of the oscillation at the highest cylinder position.

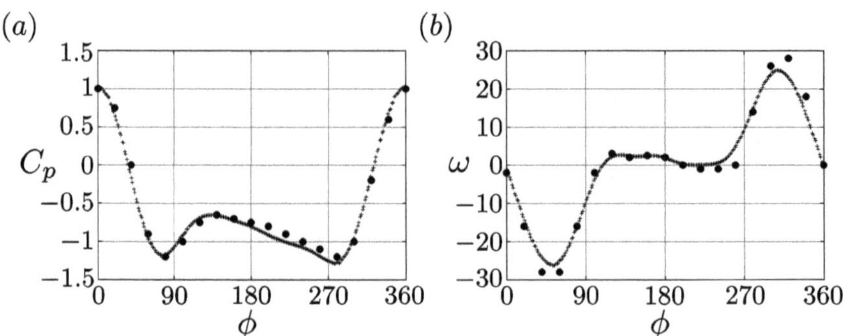

Figure 6.9 Wall quantities at the highest position of the oscillation: (a) pressure coefficient, (b) spanwise vorticity; + present results, • results by Guilmineau and Queutey (2002).

Data are compared against Guilmineau and Queutey (2002), who solved the equations over a body-fitted polar mesh fixed with the cylinder and described the flow in a moving reference frame, therefore using a conformal mesh without any approximation in the geometrical representation of the body.

These results clearly show that the present IBM can capture not only the integral values (the force coefficients) but also the local distribution of primitive and derived quantities over the immersed body surface.

7

Fluid–Structure Interaction

A common occurrence in flow dynamics is that of a body moving in a fluid in response to forces and moments exerted by, among other sources, the surrounding fluid. In this case, the motion of the body is not known a priori and it has to be computed as part of the solution, together with the flow. The latter depends on the instantaneous configuration of the boundaries, which in turn depends on the hydrodynamic loads; thus, fluid and solid systems are two-way coupled as they dynamically interact with each other.

This class of problems is referred to as fluid–structure interaction (FSI), and can involve rigid or deformable bodies; two examples are sketched in Figure 7.1 to stress the relevant differences. Rigid bodies (Figure 7.1a) have a fixed shape, so the time evolution of the center of mass position $\mathbf{X}_G$ and of the orientation (or attitude) θ is sufficient to determine the position and velocity of all boundary points and, therefore, to implement one of the IB techniques for moving interfaces described in Chapter 5.

Deformable bodies (Figure 7.1b) entail a more complex description as both the position and *shape* are unknown and, in principle, a continuum deformable surface has an infinite number of degrees of freedom. In fact, the deformable boundary can be discretized into a finite number of elements N_l whose displacement $\mathbf{d} = \{\mathbf{d}_i\}$, with $i = 1, N_l$, gives the body configuration at each time.

It is worth pointing out that for rigid bodies six unknowns (three for position and three for orientation in three dimensions) are sufficient to prescribe the configuration whose dynamics is determined by an equal number of ordinary differential equations (see Section 7.2). On the other hand, deformable bodies involve $3 \times N_l$ unknowns and their dynamics is described by a system of partial differential equations, in turn involving constitutive relations for the mechanical properties of the deforming material.

Either way, we can define a state vector $\zeta^T = \{\mathbf{q}, \mathbf{z}\}$ for the whole system comprising the fluid $\mathbf{q}^T = \{\mathbf{u}, p\}$ and the structure unknown fields $\mathbf{z}^T =$

111

$\{\mathbf{X}_G, \boldsymbol{\theta}, \mathbf{V}_G, \boldsymbol{\omega}\}$ or $\mathbf{z}^T = \{\mathbf{d}, \mathbf{v}_d\}$, respectively for rigid or deformable bodies, which are to be advanced in time to predict the system evolution. Since the structural state vector also includes velocities (the center of mass $\mathbf{V}_G$ and angular $\boldsymbol{\omega}$ velocities in the rigid case; the material point velocity of the deformable structure $\mathbf{v}_d = \{\mathbf{v}_{di}\}$), the structural problem can be written as a first-order system in time; see Section 7.1.

The integration of ζ can be achieved following different strategies, which are strongly dependent on the specific problem and essentially depend on whether fluid and structure are advanced simultaneously or sequentially. Each approach has advantages and drawbacks, and in the following we will discuss the main aspects.

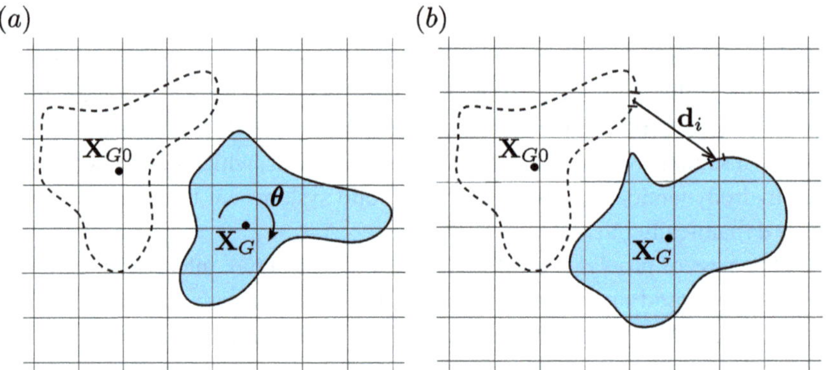

Figure 7.1 Sketch of a rigid (a) and deformable (b) body moving across an Eulerian mesh.

7.1 Coupling Strategies

Numerical algorithms for tackling FSI problems can be broadly categorized into monolithic and partitioned approaches. The former treats fluid and structure as a single dynamical system, thus solving (and moving) fluid and structure in a fully coupled fashion.

The fluid–structure operator $\mathcal{FS}$ contains both fluid equations (2.19), (2.20) and structure (see equations (7.6) or (7.29)) with the interface conditions (see equations (7.15) or (7.32), (7.33)):

$$\dot{\zeta} = \mathcal{FS}(\zeta). \tag{7.1}$$

The main advantage of monolithic solvers is their robustness, although they require a fully integrated algorithm, thus precluding the use of optimized fluid and structure solvers already available. Moreover, monolithic solvers result in a large, sparse system of nonlinear algebraic equations to be solved using iterative methods such as the Newton–Raphson, which is computationally very expensive.

Conversely, partitioned solvers treat fluid and structure as two distinct computational modules that can be solved separately, each one with its own discretization and numerical algorithm. The interfacial conditions are used explicitly to communicate information between the two solutions, possibly iteratively during each time step.

An advantage of this approach is the possibility to use available fluid and structure solvers and reduce the code development time by taking advantage of existing algorithms which have been optimized and extensively validated for a variety of problems. In compact form, the structure solver

$$\dot{\mathbf{z}} = \mathcal{S}(\mathbf{z}; \mathbf{q}) \tag{7.2}$$

embeds the structural governing equations and it depends on the instantaneous flow velocity and pressure fields through the interface condition (see equations (7.32)); hence, the fluid solution enters the structural solution through the hydrodynamic loads at the interface. On the other hand, the Navier–Stokes solver (including the corresponding boundary conditions) can be indicated in compact form:

$$\dot{\mathbf{q}} = \mathcal{F}(\mathbf{q}; \mathbf{z}), \tag{7.3}$$

where the coupling with the structural displacement field is given by the interface condition (7.15 or 7.33). Hereafter, we keep the symbolic notation for equations (7.2) and (7.3) as the details of the temporal and spatial solution depend on the specific discretization schemes and solution algorithms. For example, either implicit or explicit integration methods could be used in time while finite differences, finite volumes or finite elements could be adopted in space.

7.1.1 Loose versus Strong Coupling

In a partitioned approach, the two systems can be advanced sequentially (loose coupling) or simultaneously (strong coupling). In the former case, the fluid is solved first with the structure configuration at the previous time step (time instant t^n), then the generated hydrodynamic loads are used to evolve the structure dynamics (at time t^{n+1}). As shown in Figure 7.2a, in the loose-coupling

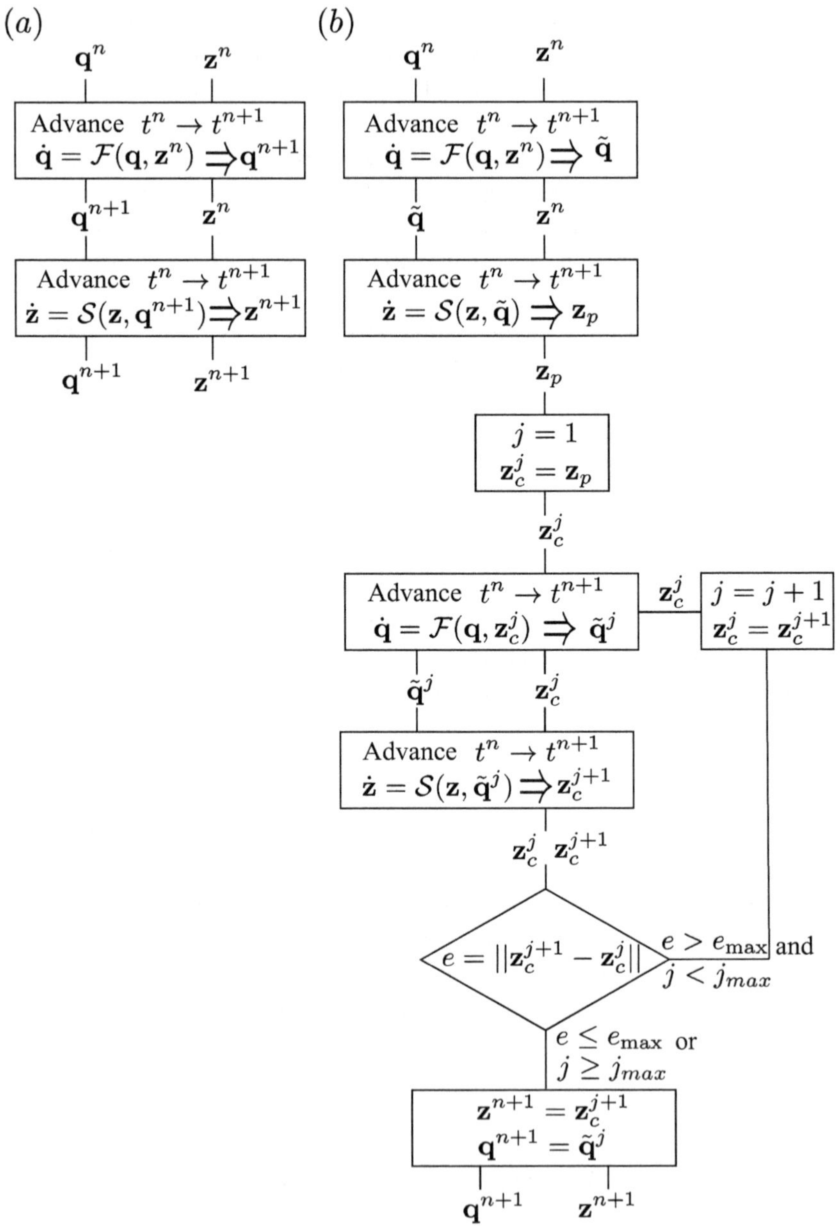

Figure 7.2 Schematic of the (a) loose and (b) strong coupling algorithms.

approach both fluid and solid solvers are invoked once per time step. Of course in the case of multistep methods, the solutions at previous time steps (e.g. $\mathbf{q}^{n-1}, \mathbf{q}^{n-2}, \ldots$ and $\mathbf{z}^{n-1}, \mathbf{z}^{n-2}, \ldots$) are also involved in the temporal scheme, but are not indicated in the sketch for the sake of clarity.

It is worth pointing out that the integration order we have just described is somehow arbitrary, since the physical problem evolves as a single phenomenon and there is no reason why the fluid should be integrated first and then the structure or vice versa. In fact, in a fluid–structure interaction problem the hydrodynamic loads (depending on the interface position and velocity) and the interface dynamics (which in turn depends on the hydrodynamic loads) should be determined simultaneously in order to correctly enforce the two-way coupling of the two systems. In the loose-coupling approach, this dynamical interdependence of fluid and solid may not be satisfied with sufficient precision at the immersed interface, thus causing numerical instability, especially in those problems with vanishing inertia of the solid phase that are therefore dominated by added mass effects (Borazjani et al., 2008).

In order to avoid this issue, in the strong coupling the governing equations for fluid and structure are solved iteratively with the interface conditions, within each time step until convergence. Figure 7.2b shows one possible workflow of the strong-coupling approach based on a predictor–corrector scheme. First, fluid and structure are solved sequentially starting from the solution at the previous time step, $\mathbf{q}^n$ and $\mathbf{z}^n$. This corresponds to a loose-coupling step yielding the *predicted* solution $\mathbf{z}_p$. Clearly, at this point the flow solution corresponding to $\mathbf{z}_p$ would be different from the intermediate solution $\tilde{\mathbf{q}}$ that was obtained using $\mathbf{z}^n$ (solving $\dot{\mathbf{q}} = \mathbf{F}(\mathbf{q}, \mathbf{z}^n)$). Hence, an iterative procedure indicated by the loop on the j index in Figure 7.2b is started. Within the loop, fluid and structure solvers are again called sequentially until the difference between the latest structural solution $\mathbf{z}_c^{j+1}$ and that of the previous iteration $\mathbf{z}_c^{j}$ is smaller than a prescribed threshold $e_{\max}$. Of course, different measures (e.g. L_∞ or L_2 norms) can be used to quantify this difference which, in turn, can be computed also on the fluid solution $\mathbf{q}$ or both $\mathbf{q}$ and $\mathbf{z}$.

Usually the most sensitive quantity is selected and this is dependent on the specific problem; however, when convergence is met the interface conditions are satisfied as the flow solution $\mathbf{q}^{n+1}$ and the corresponding hydrodynamic loads are based on the correct location and velocity of the interface, while the structural solution $\mathbf{z}^{n+1}$ is forced by the correct instantaneous hydrodynamic loads at the wet boundaries.

It is clear that in the strong coupling, the governing equations are solved implicitly through a fixed-point method, which yields a more stable and robust algorithm with respect to the loose coupling, where the hydrodynamic loads forcing the structure, as well as the position of the boundaries determining the flow, are treated explicitly.

Usually, for practical reasons, an additional condition on the maximum number of iterations $j_{\max}$ is fixed in the algorithm in order to avoid deadlocks when the error does not decrease below the prescribed threshold $e_{\max}$ after $j_{\max}$ iterations. Of course, in this case the interface conditions are not applied with the required precision and it may be necessary either to increase the maximum number of iterations $j_{\max}$ or to improve the stability of the method, for instance by decreasing the integration time step Δt or by changing the temporal scheme of the fluid or/and structural solvers.

As an example, the stability of the method can be increased by using an under-relaxation when updating the corrector step. Specifically, the corrector solution z_c^{j+1} is modified in an intermediate step before the control on the error as

$$\mathbf{z}_c^{j+1} = (1 - \xi)\tilde{\mathbf{z}}_c^{j+1} + \xi\mathbf{z}_c^{j}, \tag{7.4}$$

where $\tilde{\mathbf{z}}_c^{j+1}$ is the nonrelaxed structural solution and ξ is a constant relaxation parameter.

For large accelerations, under-relaxation can be conveniently applied to the right-hand side of the structural solver. For example, in the case of an explicit scheme for the structural solver (such as $\mathbf{z}_c^{j+1} = \mathbf{z}^n + \Delta t \mathcal{S}(\mathbf{z}_c^{j}, \tilde{\mathbf{q}}^{j})$ with explicit Euler) the equation is marched as $\mathbf{z}_c^{j+1} = \mathbf{z}^n + \Delta t[(1 - \xi)\mathcal{S}(\mathbf{z}_c^{j}, \tilde{\mathbf{q}}^{j}) + \xi\mathcal{S}(\mathbf{z}_c^{j-1}, \tilde{\mathbf{q}}^{j-1})]$. The parameter ξ is typically set to $[0.1, 0.3]$, whereas larger values further increase numerical stability but, at the same time, the rate of convergence decreases, thus requiring more subiterations and a longer computational time.

Alternatively, the rate of convergence can be increased by using accelerating techniques (Sotiropoulos and Yang, 2014) such as the Aitken method (Irons and Tuck, 1969; Mok et al., 2001), where the ξ parameter is updated dynamically during the subiterations:

$$\xi^{j+1} = 1 + (1 - \xi^j)\frac{(\Delta\mathbf{z}_c^j)^T \cdot (\Delta\mathbf{z}_c^{j+1} - \Delta\mathbf{z}_c^j)}{||\Delta\mathbf{z}_c^{j+1} - \Delta\mathbf{z}_c^j||^2}. \tag{7.5}$$

The quantity appearing on the right-hand side is defined as $\Delta\mathbf{z}_c^{j+1} = \tilde{\mathbf{z}}_c^{j+1} - \mathbf{z}_c^{j}$ and the corrected velocity is then relaxed according to equation (7.4), using the updated ξ^{j+1}. As done for the under-relaxation, the Aitken method could be applied to the structural loads (see the right-hand side of equation (7.2)).

According to this discussion, we can summarize by stating that strong coupling is more stable and robust with respect to the loose-coupling method; on the other hand, the former approach is computationally demanding since it requires us to iterate between the solvers and it is at least twice more expensive than the loose coupling as fluid and structure are solved at least twice per time step (see Figure 7.2). In contrast, loose coupling is considerably faster, but the coupling conditions at the immersed interface may not be satisfied with sufficient precision. Which of the two approaches is more efficient depends on the specific problem and its parameters, with the strong coupling adopted to prevent numerical instabilities occurring when added mass phenomena or structures with reduced inertia have to be considered.

7.2 Fluid–Structure Interaction of Rigid Bodies

The dynamics of a rigid body in a fluid–structure coupled system is governed by the balance of linear, $\mathbf{L} = M\frac{d\mathbf{X}_G}{dt}$, and angular, $\mathbf{H} = \mathbb{I}_{in}\boldsymbol{\omega}$, momentum, which in an inertial reference frame (x, y, z) read

$$M\frac{d^2\mathbf{X}_G}{dt^2} = \mathbf{F}_{in}^{ext} + \mathbf{F}_{in}^{hyd}, \qquad \frac{d\mathbb{I}_{in}\boldsymbol{\omega}}{dt} = \mathbf{M}_{in}^{ext} + \mathbf{M}_{in}^{hyd}, \qquad (7.6)$$

where M is the total mass of the body (here assumed not varying in time) and $\mathbf{X}_G$ the instantaneous position of the center of mass. The angular velocity of the body $\boldsymbol{\omega}$ is, in general, not parallel to the angular momentum owing to the action of the inertia tensor $\mathbb{I}_{in}$, where the *in* suffix indicates that the tensor is expressed with respect to the inertial frame. $\mathbf{F}_{in}^{ext}$ gathers all other external forces acting on the body (e.g. gravitational force, springs and dashpots forces) and $\mathbf{M}_{in}^{ext}$ is the corresponding moment, computed with respect to a convenient point. $\mathbf{F}_{in}^{hyd}$ and $\mathbf{M}_{in}^{hyd}$ are the hydrodynamic force and moment computed in the inertial frame:

$$\mathbf{F}_{in}^{hyd} = \int_S (-p\mathbf{n} + \boldsymbol{\tau} \cdot \mathbf{n})dS, \qquad \mathbf{M}_{in}^{hyd} = \int_S \mathbf{X} \times (-p\mathbf{n} + \boldsymbol{\tau} \cdot \mathbf{n})dS, \qquad (7.7)$$

with pressure p and viscous stress tensor $\boldsymbol{\tau}$ depending on the local position $\mathbf{X}$ over the wet surface S.

Solving the balance of angular momentum in an inertial reference frame can be impractical as both $\mathbb{I}_{in}$ and $\boldsymbol{\omega}$ vary in time for a general motion of the rigid body. In such cases, it is more convenient to write the second of equations (7.6) in a body-fixed reference frame (x^*, y^*, z^*), since in this noninertial frame the inertia tensor is time independent (unless the body is losing or gaining mass, as discussed in Section 9.4). The equations become:

$$M\frac{d^2\mathbf{X}_G}{dt^2} = \mathbf{F}_{in}^{ext} + \mathbf{F}_{in}^{hyd}, \qquad \mathbb{I}_*\dot{\boldsymbol{\omega}} + \boldsymbol{\omega} \times \mathbb{I}_*\boldsymbol{\omega} = \mathbf{M}_*^{ext} + \mathbf{M}_*^{hyd}, \qquad (7.8)$$

where $\mathbb{I}_*$ is the inertia tensor with respect to (x^*, y^*, z^*) and the right-hand side of the second equation comprises the moments in the noninertial body frame, with

$$\mathbf{M}_*^{hyd} = \int_S \mathbf{x} \times (-p\mathbf{n} + \boldsymbol{\tau} \cdot \mathbf{n})dS, \qquad (7.9)$$

where $\mathbf{x} = \mathbf{X} - \mathbf{X}_G$. The external moment in the body-fixed coordinate system (7.9) can also be obtained directly from the one expressed in the laboratory coordinate system (7.7). If the origins of the two frames are coincident, a rotation matrix $\mathbb{R}$ (detailed at the end of this section) links the two through

$\mathbf{M}^{ext} = \mathbb{R}\mathbf{M}_*^{ext}$ (the same expression holds true for the hydrodynamic moments).

Furthermore, when choosing the principal axes of inertia as the body frame, the inertia tensor becomes diagonal $\mathbb{I}_*$ and the second of equations (7.8) reduces to Euler's equations, whose components read:

$$I_{*,x^*}\dot{\omega}_{x^*} - (I_{*,y^*} - I_{*,z^*})\omega_{y^*}\omega_{z^*} = M_{*,x^*}^{ext} + M_{*,x^*}^{hyd},$$

$$I_{*,y^*}\dot{\omega}_{y^*} - (I_{*,z^*} - I_{*,x^*})\omega_{z^*}\omega_{x^*} = M_{*,y^*}^{ext} + M_{*,y^*}^{hyd}, \qquad (7.10)$$

$$I_{*,z^*}\dot{\omega}_{z^*} - (I_{*,x^*} - I_{*,y^*})\omega_{x^*}\omega_{y^*} = M_{*,z^*}^{ext} + M_{*,z^*}^{hyd},$$

where $I_{*,x^*}, I_{*,y^*}, I_{*,z^*}$ are the principal moments of inertia. Euler's equations (7.10) form a nonlinear system of equations that can be integrated to obtain the angular velocity components in the rotating body-fixed frame $\omega_{x^*}, \omega_{y^*}, \omega_{z^*}$.

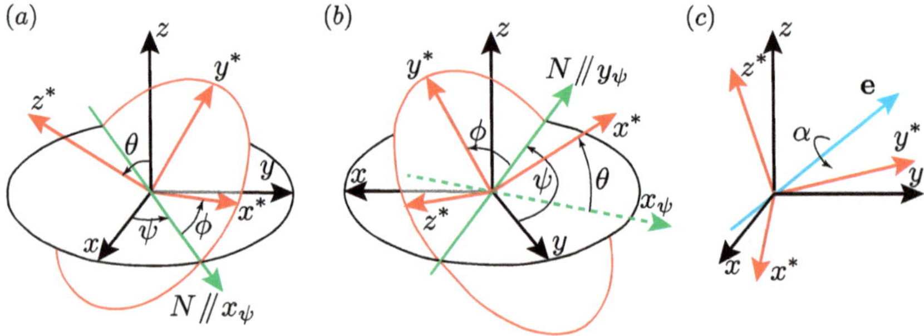

Figure 7.3 Geometrical definition of the (a) Euler, (b) Euler–Cardan angles and (c) quaternion. The solid green line in (a) and (b) is the line of nodes, which is aligned respectively with x and y after the first rotation around z. The dashed line in (b) indicates the orientation of the x-axis after the first rotation of angle ψ around z. The blue solid line in (c) indicates the Euler eigenaxis; α is the corresponding rotation angle to align the two frames.

The instantaneous position of the body with respect to the fixed inertial coordinate frame can be obtained by expressing the components of $\boldsymbol{\omega}$ in the rotating frame as a function of the Euler angles and of their derivatives (Goldstein et al., 2001; Wie, 1998):

$$\begin{pmatrix} \dot{\psi} \\ \dot{\theta} \\ \dot{\phi} \end{pmatrix} = \begin{bmatrix} \sin\phi/\sin\theta & \cos\phi/\sin\theta & 0 \\ \cos\phi & -\sin\phi & 0 \\ -\sin\phi/\tan\theta & -\cos\phi/\tan\theta & 1 \end{bmatrix} \begin{pmatrix} \omega_{x^*} \\ \omega_{y^*} \\ \omega_{z^*} \end{pmatrix}, \qquad (7.11)$$

where (ψ, θ, ϕ) are the precession, nutation and intrinsic rotation or spin (see Figure 7.3a). We recall that, in order to align a fixed inertial frame (x, y, z) with the body rotating frame (x^*, y^*, z^*) using the sequence of Euler's angles, first a rotation of ψ about z is applied, such that the x-axis is aligned with the direction orthogonal to both z and z^* (the line of nodes); secondly, a rotation of θ about the line of nodes is applied so the z-axis becomes aligned with z^*; lastly, a rotation of ϕ about the transformed z (now aligned with z^*) is applied, such that the transformed x and y become aligned with x^* and y^*, respectively.

Equations (7.11) can be integrated to obtain the evolution of the Euler angles and determine the instantaneous orientation of the body with respect to the fixed inertial frame. Similarly, these equations can be injected in Euler's equations (7.10), thus yielding a second-order nonlinear system in $(\ddot{\psi}, \ddot{\theta}, \ddot{\phi})$.

The relation (7.11) between the time rate of change of the Euler angles and the angular velocity ω also shows that Euler angles are singular when the nutation angle approaches 0 or π, with the spin and precession angular velocity diverging. In such cases when $\theta \approx 0, \pi$ is not prevented by a constraint on the rigid body, a different rule to describe the orientation of the rigid body with respect to the fixed coordinate system should be used. Indeed, other sequences of rotations are possible to align the instantaneous body frame with the inertial one. As an example, the Euler–Cardan angles (also known as Tait–Bryan angles (Goldstein et al., 2001)) are used in aeronautics to describe the orientation of an aircraft with respect to a fixed frame through yaw ψ, pitch θ and roll ϕ angles, which implies a different definition for the line of nodes. As shown in Figure 7.3b, in order to align a fixed inertial frame (x, y, z) with the body rotating frame (x^*, y^*, z^*) first a rotation of ψ about z is applied, such that the y-axis is aligned with direction orthogonal to both z and x^* (the lines of nodes); second, a rotation of θ about the lines of nodes is performed, so the x-axis becomes aligned with x^*; last, a rotation of ϕ about the transformed x (now aligned with x^*) is applied, such that the transformed y and z align with y^* and z^*, respectively. The corresponding relation between the rotation rate of the Euler–Cardan angles and the angular velocity components (expressed in the moving frame) is

$$
\begin{pmatrix} \dot{\psi} \\ \dot{\theta} \\ \dot{\phi} \end{pmatrix} = \begin{bmatrix} 0 & \sin\phi/\cos\theta & \cos\phi/\cos\theta \\ 0 & \cos\phi & -\sin\phi \\ 1 & \tan\theta\sin\phi & \tan\theta\cos\phi \end{bmatrix} \begin{pmatrix} \omega_{x^*} \\ \omega_{y^*} \\ \omega_{z^*} \end{pmatrix}. \tag{7.12}
$$

Note that this relation can also become singular for $\theta = \pm\pi/2$, with the rate of change of roll and yaw angles diverging. For a general rotational motion of a rigid body that may encounter the singularity of the Euler angles, it is then more suitable to resort to the method of quaternions to keep track of the instantaneous orientation of the body (Wie, 1998).

The components of the quaternion corresponding to the instantaneous body orientation are $q_0 = \cos\alpha/2$, $q_1 = e_1 \sin\alpha/2$, $q_2 = e_2 \sin\alpha/2$, $q_3 = e_3 \sin\alpha/2$, where $\mathbf{e} = (e_1, e_2, e_3)$ are the direction cosines of Euler's eigenaxis, which is fixed in both the inertial (x, y, z) and body (x^*, y^*, z^*) reference frames (and thus the direction cosines are the same in the two frames). The angle α is the rotation angle about $\mathbf{e}$ to rotate the inertial frame onto the body frame; see Figure 7.3c. The kinematic differential equations for the quaternion are written as follows:

$$
\begin{pmatrix} \dot{q}_0 \\ \dot{q}_1 \\ \dot{q}_2 \\ \dot{q}_3 \end{pmatrix} = \frac{1}{2} \begin{bmatrix} -q_1 & -q_2 & -q_3 \\ q_0 & -q_3 & q_2 \\ q_3 & q_0 & -q_1 \\ -q_2 & q_1 & q_0 \end{bmatrix} \begin{pmatrix} \omega_{x^*} \\ \omega_{y^*} \\ \omega_{z^*} \end{pmatrix},
\tag{7.13}
$$

or, alternatively,

$$
\begin{pmatrix} \dot{q}_0 \\ \dot{q}_1 \\ \dot{q}_2 \\ \dot{q}_3 \end{pmatrix} = \frac{1}{2} \begin{bmatrix} 0 & -\omega_{x^*} & -\omega_{y^*} & -\omega_{z^*} \\ \omega_{x^*} & 0 & \omega_{z^*} & -\omega_{y^*} \\ \omega_{y^*} & -\omega_{z^*} & 0 & \omega_{x^*} \\ \omega_{z^*} & \omega_{y^*} & -\omega_{x^*} & 0 \end{bmatrix} \begin{pmatrix} q_0 \\ q_1 \\ q_2 \\ q_3 \end{pmatrix},
\tag{7.14}
$$

which can be written in more compact form as $\dot{\mathbf{q}} = \frac{1}{2}(q_0\boldsymbol{\omega} - \boldsymbol{\omega} \times \mathbf{q})$ and $\dot{q}_0 = -\frac{1}{2}\boldsymbol{\omega} \cdot \mathbf{q}$, with $\mathbf{q} = (q_1, q_2, q_3)$ and $\mathbf{q} \cdot \mathbf{q} + q_0^2 = 1$.

Lastly, the resulting body dynamics has to be used to impose the no-slip (or whatever) boundary condition through the IBM at the fluid–solid interface. At any wet point, the body velocity $\mathbf{U}$ in the inertial reference frame reads

$$
\mathbf{U} = \frac{\mathrm{d}\mathbf{X}_G}{\mathrm{d}t} + (\mathbb{R}\boldsymbol{\omega}) \times (\mathbf{X} - \mathbf{X}_G),
\tag{7.15}
$$

where $\mathbf{X}$ and $\mathbf{X}_G$ are the vector positions of the wet point and of the center of mass in the inertial frame, respectively. The angular velocity $\boldsymbol{\omega}$ comes from the solution of Euler's equations (7.10) and is thus expressed in the noninertial body frame. The rotation matrix $\mathbb{R}$ provides the transformation from the body-fixed frame to the inertial frame. Hence, the first term on the right-hand side of equation (7.15) is the translation velocity of the center of mass of the rigid body and the second one is the rotation counterpart, both expressed in the laboratory frame.

When the body orientation is described using Euler angles, the rotation matrix reads

$$
\mathbb{R} = \begin{bmatrix} \cos\psi\cos\phi - \cos\theta\sin\psi\sin\phi & -\cos\psi\sin\phi - \cos\theta\sin\psi\cos\phi & \sin\theta\sin\psi \\ \sin\psi\cos\phi + \cos\theta\cos\psi\sin\phi & -\sin\psi\sin\phi + \cos\theta\cos\psi\cos\phi & -\sin\theta\cos\psi \\ \sin\theta\sin\phi & \sin\theta\cos\phi & \cos\theta \end{bmatrix},
\tag{7.16}
$$

whereas for the Euler–Cardan angles is equal to

$$
\mathbb{R} = \begin{bmatrix} \cos\theta\cos\psi & \sin\phi\sin\theta\cos\psi - \cos\phi\sin\psi & \cos\phi\sin\theta\cos\psi + \sin\phi\sin\psi \\ \cos\theta\sin\psi & \sin\phi\sin\theta\sin\psi + \cos\phi\cos\psi & \cos\phi\sin\theta\sin\psi - \sin\phi\cos\psi \\ -\sin\theta & \sin\phi\cos\theta & \cos\phi\cos\theta \end{bmatrix},
$$
$$(7.17)$$

and for the quaternions

$$
\mathbb{R} = \begin{bmatrix} 1 - 2(q_2^2 + q_3^2) & 2(q_1 q_2 - q_3 q_0) & 2(q_1 q_3 + q_2 q_0) \\ 2(q_1 q_2 + q_3 q_0) & 1 - 2(q_1^2 + q_3^2) & 2(q_2 q_3 - q_1 q_0) \\ 2(q_1 q_3 - q_2 q_0) & 2(q_2 q_3 + q_1 q_0) & 1 - 2(q_1^2 + q_2^2) \end{bmatrix}.
\qquad (7.18)
$$

7.3 Fluid–Structure Interaction of Deformable Bodies

We consider now the problem in which the fluid interacts with a deformable body. In this case, determining the position of the center of mass and the orientation of the body is not enough to characterize its dynamics, as the body changes shape during the motion. Moreover, the instantaneous configuration of the boundary not only depends on the local hydrodynamic loads acting on its wet surface(s), but also on the mechanical properties of the solid material or, more precisely, on its constitutive relation.

Before tackling the fluid–structure interaction problem, we briefly recall the governing structural equations for an isothermal material. A comprehensive account of available constitutive relations, also including mechanical and thermal balance laws, can be found in dedicated textbooks (Gonzalez and Stuart, 2008; Holzapfel, 2002; Tadmor et al., 2012).

As shown in Figure 7.4, the motion of a deformable continuum Ω can be described relative to an arbitrary reference state, which may be the configuration at time $t = 0$ (initial configuration) or that in which the material is stress-free (undeformed configuration).

In a Cartesian frame, a body point P (or material particle) in the reference configuration can then be assigned a unique position vector $\mathbf{X}$ and its coordinates X_1, X_2, X_3 are referred to as material (or Lagrangian) coordinates. At a successive time, say t, the body attains a different configuration (current or deformed configuration) and the same particle P now occupies the location $\mathbf{x}$, with spatial (or Eulerian) coordinates x_1, x_2, x_3. Hence, each material particle has two sets of coordinates associated with it, namely the material coordinates that stay with it throughout its motion and spatial coordinates which change as the body moves and deforms. The motion of any point can then be

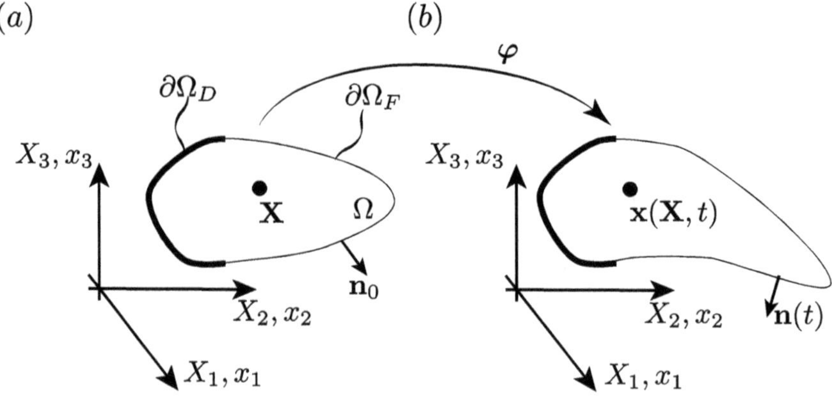

Figure 7.4 (a) Reference and (b) current configuration of a deformable body with displacement and traction conditions on $\partial\Omega_D$ and $\partial\Omega_F$, which are the boundaries of the closed material domain Ω.

expressed as a relationship between material and spatial coordinates. In the material description, the instantaneous position of the material particle can be determined from

$$\mathbf{x} = \varphi(\mathbf{X}, t). \tag{7.19}$$

Alternatively, if one knows the current position, then the material coordinate can be determined from the inverse relation

$$\mathbf{X} = \varphi^{-1}(\mathbf{x}, t), \tag{7.20}$$

where φ is the deformation field that maps each material point from the reference to the deformed configuration.[1] Any physical property of the body (for example density and temperature) or kinematic property (such as displacement or velocity) can thus be described in terms of either the material $\mathbf{X}$ or spatial $\mathbf{x}$ coordinates, since they can be transformed into each other using equations (7.19) and (7.20).

---

[1] The function φ must satisfy some properties to be *admissible* such as being bijective for the material and deformed configurations. Furthermore, it has to be always true that $\det[\nabla\varphi] > 0$, which has to do with a deformed configuration with local positive volume (Gonzalez and Stuart, 2008).

7.3.1 Structural Equations in Eulerian Form

The difference between the instantaneous position of each material particle of the body, $\mathbf{x}$, and its counterpart in the reference configuration $\mathbf{X}(\mathbf{x}, t) = \varphi^{-1}(\mathbf{x}, t)$ defines the spatial displacement field $\mathbf{d}(\mathbf{x}, t) = \mathbf{x} - \mathbf{X}(\mathbf{x}, t)$. The equation of motion for a deformable body in the spatial (or Eulerian) form can then be written as

$$\rho_s \frac{\mathrm{d}^2\mathbf{d}}{\mathrm{d}t^2} = \nabla \cdot \mathbb{T} + \rho_s \mathbf{b}, \tag{7.21}$$

where the inertial term on the left-hand side is the local momentum variation given by the material time derivative of the displacement field. Such an unsteady term can also be written in terms of the displacement velocity

$$\mathbf{v}_d(\mathbf{x}, t) = \frac{\mathrm{d}\,\mathbf{d}(\mathbf{x}, t)}{\mathrm{d}t}, \quad \text{as} \quad \frac{\mathrm{d}^2\mathbf{d}}{\mathrm{d}t^2} = \frac{\mathrm{d}\mathbf{v}_d(\mathbf{x}, t)}{\mathrm{d}t} = \frac{\partial \mathbf{v}_d}{\partial t} + \mathbf{v}_d \cdot \nabla \mathbf{v}_d,$$

where $\nabla \mathbf{v}_d$ is the gradient of the velocity of a solid particle when moving in space, so that the equation reads

$$\rho_s \frac{\mathrm{d}\mathbf{v}_d}{\mathrm{d}t} = \nabla \cdot \mathbb{T} + \rho_s \mathbf{b}. \tag{7.22}$$

Here, $\rho_s(\mathbf{x}, t)$ is the spatial mass density, which is a continuous function of the Eulerian coordinates. Its spatial and temporal evolution is governed by the equation of mass conservation

$$\frac{\mathrm{d}\rho_s}{\mathrm{d}t} + \rho_s \nabla \cdot \mathbf{v}_d = 0. \tag{7.23}$$

The Cauchy stress tensor $\mathbb{T}$ provides the local traction $\mathbf{t} = \mathbb{T} \cdot \mathbf{n}$, which represents the force per unit area acting on a surface element in the *current* configuration, oriented with outward normal vector $\mathbf{n}$. Since the Cauchy stress measures the force per unit area in the current, deformed configuration, it is also called the *true* stress.

As for the fluid phase, the balance of angular momentum is assured by the symmetry of the Cauchy stress:

$$\mathbb{T} = \mathbb{T}^T. \tag{7.24}$$

The last term on the right-hand side of equation (7.21) (and (7.22)) contains all volume forces (per unit mass) such as those due to a gravity field or centrifugal forces induced by background rotation, when using the noninertial reference frame.

The governing equations should be completed by boundary and initial conditions. The latter requires the imposition of the initial displacement and its velocity for the solid continuum:

$$\mathbf{d}(\mathbf{x}, t) = \mathbf{d}_0(\mathbf{x}) \quad \text{at } t = 0,$$
$$\dot{\mathbf{d}}(\mathbf{x}, t) = \dot{\mathbf{d}}_0(\mathbf{x}) \quad \text{at } t = 0. \tag{7.25}$$

Usually there are two types of boundary conditions, those on displacement and those on traction (Figure 7.4). The former is a Dirichlet type condition setting the position of material particles over some portion of the boundary: $\mathbf{d} = \bar{\mathbf{d}}$ on $\partial\Omega_D$. As an example, if a portion of the boundary is fixed to the initial position, then $\mathbf{d} = \mathbf{0}$ on that portion. On the other hand, the traction boundary condition specifies the traction $\mathbf{t}$ over the remaining portion of the body, $\partial\Omega_F$. This is the kind of boundary condition to be imposed at the wet surface of the body, where the prescribed traction is given by the instantaneous hydrodynamic stress

$$\mathbf{t} = \mathbb{T} \cdot \mathbf{n} = -p\mathbf{n} + \tau \cdot \mathbf{n} \quad \text{on } \partial\Omega_F, \tag{7.26}$$

where $\mathbf{n}$ is the outward normal from the solid to the fluid, p is the pressure and τ is the viscous stress tensor of the fluid (see equation (6.1)). Additionally, the Navier–Stokes equations (2.19) and (2.20) are coupled to the solid continuum through compatibility relations, such as the no-slip condition at the fluid/solid boundary:

$$\mathbf{u}(\mathbf{x}, t) = \dot{\mathbf{d}}(\mathbf{x}, t). \tag{7.27}$$

7.3.2 Structural Equations in Lagrangian Form

The governing equations for structural mechanics can be written, in a more convenient form, in terms of Lagrangian variables. Indeed, within this description, stress and strain are expressed using the fixed reference configuration, where the independent variables are Lagrangian (material) coordinates $\mathbf{X}$ and the time t. The momentum equation has a similar form to the Eulerian description:

$$\rho_{s0}\frac{\partial^2 \mathbf{d}}{\partial t^2} = \nabla \cdot \mathbb{P} + \rho_{s0}\mathbf{b}, \tag{7.28}$$

where ρ_{s0} is the initial density field of the body (in the reference configuration) and $\mathbf{d}(\mathbf{X}, t) = \varphi(\mathbf{X}, t) - \mathbf{X}$ is the displacement field expressed with the material coordinates. The applied loads are defined on the reference configuration through the first Piola–Kirchhoff stress tensor $\mathbb{P}$, which measures the force in the actual configuration but normalized per unit surface area in the reference configuration. Alternatively, the momentum equation (7.28) can be rewritten in terms of the material point velocity as

$$\rho_{s0}\frac{\partial \mathbf{v}_d}{\partial t} = \nabla \cdot \mathbb{P} + \rho_{s0}\mathbf{b}, \quad \text{with} \quad \mathbf{v}_d = \frac{\partial \mathbf{d}}{\partial t}. \tag{7.29}$$

The first Piola–Kirchhoff stress tensor $\mathbb{P}$ is related to the Cauchy stress through $\mathbb{P} = J\mathbb{T}\mathbb{F}^{-T}$, where $J = \det[\mathbb{F}]$ is the determinant of the deformation gradient $\mathbb{F} \equiv \nabla\varphi = \partial\mathbf{x}/\partial\mathbf{X}$. Hence, the balance of the angular momentum is given by

$$\mathbb{P}\mathbb{F}^{T} = \mathbb{F}\mathbb{P}^{T}. \tag{7.30}$$

It should be noted that only the density field in the reference configuration, ρ_{s0}, enters the momentum equation (7.28) (or (7.29)), so the continuity equation in Lagrangian form can be written in terms of reference ρ_{s0} and spatial ρ_s mass density by requiring the infinitesimal mass dm at position $\mathbf{X}$ in the reference configuration to be the same as the differential mass in the current configuration $\mathbf{x}$: $dm = \rho_{s0}(\mathbf{X})dV = \rho_s(\mathbf{x}, t)dv$. Recalling that the reference and instantaneous differential volume are related by $dv = JdV$, the mass conservation reads

$$\rho_{s0}(\mathbf{X}) = \rho_s(\mathbf{x}, t)J(\mathbf{X}, t). \tag{7.31}$$

The initial conditions are the same as (7.25), but expressed in Lagrangian coordinates. Similarly, the boundary condition on Ω_D reads $\mathbf{d}(\mathbf{X}) = \overline{\mathbf{d}}$, whereas the traction boundary condition on the wet surface of the body, $\partial\Omega_F$, is obtained by converting the traction exerted by the fluid on the instantaneous configuration to the reference configuration

$$\mathbb{P} \cdot \mathbf{n}_0 = J(-p\mathbf{I} + \tau)\mathbb{F}^{-T} \cdot \mathbf{n}_0 \quad \text{on } \partial\Omega_F, \tag{7.32}$$

where $\mathbf{n}_0$ is the local normal vector of the reference configuration and $\mathbf{I}$ is the identity tensor. The no-slip condition in the Navier–Stokes equations on the wet solid boundary is given by

$$\mathbf{u}(\mathbf{x}, t) = \dot{\mathbf{d}}(\mathbf{X}, t). \tag{7.33}$$

7.3.3 Constitutive Relations

In the Eulerian description, the structural governing equations have 13 unknown scalar fields, namely the spatial density $\rho_s(\mathbf{x})$, 3 components of the displacement $\mathbf{d}(\mathbf{x}, t)$ and 9 elements of the Cauchy stress tensor $\mathbb{T}(\mathbf{x}, t)$. To determine these unknown fields we have so far written 7 independent equations given by the linear (7.21) and angular (7.24) vector equations, along with the conservation of mass (7.23). On the other hand, in the Lagrangian description only the reference density ρ_{s0} (which is known) appears in the momentum equation.

In this setting, we thus have 12 unknown scalar fields – 3 components of the displacement field $\mathbf{d}(\mathbf{X}, t)$ and 9 stress components of the first Piola–Kirchhoff stress tensor $\mathbb{P}(\mathbf{x}, t)$ – and 6 independent equations (7.28) and (7.30). Indeed, the equation of mass conservation is decoupled as it is only needed to determine the instantaneous density field. In both cases, therefore, 6 additional equations are required, which are provided by the constitutive relations. The latter are an intrinsic property of the continuum material, relating the independent components of the stress tensor to the strain at any point and time.

For Cauchy-elastic materials, the stress field depends only on the *instantaneous* deformation and thus it does not depend on the deformation history. However, the actual work done by the stress field and the energy of the material could depend on the deformation path. The constitutive equation of an isothermal elastic body relates the Cauchy stress tensor $\mathbb{T}(\mathbf{x}, t)$ to the deformation gradient $\mathbb{F}(\mathbf{X}, t)$:

$$\mathbb{T}(\mathbf{x}, t) = g_C(\mathbb{F}), \tag{7.34}$$

where g_C is the tensor-valued constitutive relation. Alternatively, the constitutive relation can be written as

$$\mathbb{P}(\mathbf{x}, t) = g_{PK1}(\mathbb{F}), \qquad \mathbb{S}(\mathbf{x}, t) = g_{PK2}(\mathbb{F}), \tag{7.35}$$

where g_{PK1} and g_{PK2} are the tensor-valued constitutive relations associated with the first $\mathbb{P} = J\mathbb{T}\mathbb{F}^{-T}$ or second $\mathbb{S} = J\mathbb{F}^{-1}\mathbb{T}\mathbb{F}^{-T} = \mathbb{F}^{-1}\mathbb{P}$ Piola–Kirchhoff stress tensors.

Cauchy-elastic materials, for which the work done is independent of the load path, are referred to as *hyperelastic* and the corresponding stress can be derived from a stored strain energy potential ψ. The constitutive relation for a hyperelastic material based on the first Piola–Kirchhoff stress tensor then reads

$$\mathbb{P}(\mathbf{x}, t) = g_{PK1}(\mathbb{F}) = \frac{\partial \psi(\mathbb{F})}{\partial \mathbb{F}}. \tag{7.36}$$

Alternatively, the second Piola–Kirchhoff stress can be expressed as a function of other strain measures such as the right Cauchy–Green deformation tensor $\mathbb{C} = \mathbb{F}^T\mathbb{F}$ or the Green–Lagrange deformation tensor $\mathbb{E} = (\mathbb{C} - \mathbb{I})/2$:

$$\mathbb{S}(\mathbf{x}, t) = 2\frac{\partial \psi(\mathbb{C})}{\partial \mathbb{C}}, \qquad \mathbb{S}(\mathbf{x}, t) = \frac{\partial \psi(\mathbb{E})}{\partial \mathbb{E}}, \tag{7.37}$$

where $\mathbb{I}$ is the identity matrix. Since the strain energy and its gradient should be frame invariant, for an isotropic continuum the deformation energy $\psi(I_1, I_2, I_3)$ depends only on the principal invariants of the right Cauchy–Green tensor $\mathbb{C}$, namely, $I_1 = \text{tr}(\mathbb{C})$, $I_2 = (\text{tr}^2(\mathbb{C}) - \text{tr}(\mathbb{C}^2))/2$, $I_3 = \det(\mathbb{C}) = J^2$. Consequently, according to the second of equations (7.37), the second Piola–Kirchhoff stress tensor is obtained as

$$\mathbb{S} = 2\frac{\partial \psi}{\partial \mathbb{C}} = 2\frac{\partial \psi}{\partial I_1}\frac{\partial I_1}{\partial \mathbb{C}} + 2\frac{\partial \psi}{\partial I_2}\frac{\partial I_2}{\partial \mathbb{C}} + 2\frac{\partial \psi}{\partial I_3}\frac{\partial I_3}{\partial \mathbb{C}}. \tag{7.38}$$

As an example, for incompressible materials the neo-Hookean constitutive model depends on the first invariant

$$\psi(\mathbb{C}) = c_1(I_1 - 3), \tag{7.39}$$

whereas the Mooney–Rivlin model depends on the first and second invariant

$$\psi(\mathbb{C}) = c_1(I_1 - 3) + c_2(I_2 - 3), \tag{7.40}$$

where c_1 and c_2 are material constants. Another example of a hyperelastic constitutive relation is the Saint Venant–Kirchhoff model:

$$\psi(\mathbb{E}) = \frac{\lambda}{2}\mathrm{tr}^2(\mathbb{E}) + \mu\mathbb{E} : \mathbb{E}, \tag{7.41}$$

which results in the second Piola–Kirchhoff stress $\mathbb{S} = \lambda\mathrm{tr}(\mathbb{E})\mathbb{I} + 2\mu\mathbb{E}$, where λ and μ are the Lamé coefficients. It should be noted that although the constitutive relations result in a linear relation in between $\mathbb{S}$ and $\mathbb{E}$, the constitutive model is nonlinear as the Green–Lagrange strain depends in a nonlinear way on the deformation field.

For small deformations, the Green–Lagrange tensor can be replaced by the small strain tensor $\mathbb{e} = (\nabla\mathbf{d} + (\nabla\mathbf{d})^T)/2$ (since $\mathbb{E} = \mathbb{e} + O((\nabla\mathbf{d})^2)$ for $||\nabla\mathbf{d}|| \to 0$) and there is no distinction to be made between this deformed configuration and some reference, or undeformed, configuration. As a consequence, $\mathbb{T} = \mathbb{P} + O(\nabla\mathbf{d}) = \mathbb{S} + O(\nabla\mathbf{d})$ and the linear constitutive relation can be written with respect to the Cauchy stress tensor:

$$\mathbb{T} = \lambda\nabla \cdot \mathbf{d}\mathbb{I} + 2\mu\mathbb{e}. \tag{7.42}$$

Therefore, the Navier–Lamé momentum equation for a linear and isotropic material is

$$\rho_s\frac{\partial^2 \mathbf{d}}{\partial t^2} = (\lambda + \mu)\nabla(\nabla \cdot \mathbf{d}) + \mu\nabla^2\mathbf{d} + \rho_s\mathbf{b}, \tag{7.43}$$

with ρ_s the material density and $\mathbf{b}$ the body mass force.

Biological Tissues

So far, we have limited the discussion to isotropic materials, which retain the same elastic properties in all directions. Many relevant materials, however, are anisotropic, meaning their deformation depends on the direction of the applied loads. An important example is the biological tissues, which are made of bundles of parallel fibers, in turn organized into parallel sheets bonded across the tissue thickness; these structures are often modeled as orthotropic materials characterized by three orthogonal planes of material symmetry with unique and independent mechanical properties along each direction.

As an example, the passive elasticity of the myocardial tissue of the human heart is often modeled through an orthotropic constitutive relation using a Fung-type exponential strain-energy function (Usyk et al., 2000; Viola et al., 2023b):

$$\psi(\mathbb{E}) = C(e^Q - 1)/2 + C_{compr}(J \ln J - J + 1), \quad \text{with}$$
$$Q = b_{ff}E_{ff}^2 + b_{ss}E_{ss}^2 + b_{nn}E_{nn}^2 + 2b_{fs}E_{fs}^2 + 2b_{fn}E_{fn}^2 + 2b_{ns}E_{ns}^2, \tag{7.44}$$

where E_{ij} are the components of the strain tensor $\mathbb{E}$ in an orthogonal coordinate system having fiber, sheet and sheet-normal $(\hat{\mathbf{f}}, \hat{\mathbf{s}}, \hat{\mathbf{n}})$ axes respectively. The model has eight material parameters, where the last term in the first of equations (7.44) allows tissue compressibility, whereas large C_{compr} yields nearly incompressible tissue.

7.4 Internal Structural Stresses

The resulting governing equations for the structure dynamics, completed with initial and boundary conditions, have to be solved numerically. In this section we briefly recall the traditional finite-element method (FEM) applied to a linear elastic continuum. Then, as a possible alternative, the discrete method of interaction potentials is briefly described.

7.4.1 Finite-Element Method

Let us consider the linear elasticity problem over a closed solid domain Ω; see Figure 7.4. Its boundary $\partial\Omega$ is partially fixed $\partial\Omega_D$ (homogeneous Dirichlet condition) and the rest $\partial\Omega_F$ (traction condition) is wet by the surrounding fluid:

$$\begin{cases} \rho_s \frac{\partial^2 \mathbf{d}}{\partial t^2} = \nabla \cdot \mathbb{T} + \rho_s \mathbf{b}, \\ \mathbf{d}|_{\partial\Omega_D} = \mathbf{0}, \\ \mathbb{T} \cdot \mathbf{n}|_{\partial\Omega_F} = \mathbf{t}_{\text{fluid}}, \end{cases} \tag{7.45}$$

where we recall that $\mathbf{d}$ is the unknown displacement field and $\mathbf{t}_{\text{fluid}} = -p\mathbf{n} + \tau \cdot \mathbf{n}$ is the traction force exerted from the fluids on the wet boundary (see Section

7.3.1). The system is completed with initial conditions on the solid displacement and velocity. The problem is then projecting equation (7.45) over a suitable test function ζ, whose components belong to the same space of the solution $\mathbf{d}$. Hence, assuming the domain and the boundary conditions to be sufficiently regular, the following weak form is obtained:

$$\int_\Omega \rho_s \frac{\partial^2 \mathbf{d}}{\partial t^2} \cdot \zeta \mathrm{d}\Omega = \int_\Omega (\nabla \cdot \mathbb{T}) \cdot \zeta \mathrm{d}\Omega + \int_\Omega \rho_s \mathbf{b} \cdot \zeta \mathrm{d}\Omega. \tag{7.46}$$

Using the Green formula and the divergence theorem, the first term on the right-hand side can be rewritten as

$$\int_\Omega (\nabla \cdot \mathbb{T}) \cdot \zeta \mathrm{d}\Omega = \underbrace{\oint_{\partial\Omega} (\mathbb{T} \cdot \boldsymbol{n}) \cdot \zeta \mathrm{d}\gamma}_{\int_\Omega \nabla \cdot (\mathbb{T}\zeta) \mathrm{d}\Omega} - \int_\Omega \mathbb{T} : \nabla\zeta \mathrm{d}\Omega. \tag{7.47}$$

The boundary integral contains the scalar product of the test function and the traction over the surface of the tissue. Consequently, since $\zeta = 0$ on $\partial\Omega_D$, the boundary integral is zero over the Dirichlet boundary, whereas on the wet boundary $\int_{\partial\Omega_F} (\mathbb{T} \cdot \boldsymbol{n}) \cdot \zeta \mathrm{d}\gamma = \int_{\partial\Omega_F} \mathbf{t}_{\text{fluid}} \cdot \zeta \mathrm{d}\gamma$. Hence, the weak form of the governing equations and boundary conditions reads

$$\int_\Omega \rho_s \frac{\partial^2 \mathbf{d}}{\partial t^2} \cdot \zeta \mathrm{d}\Omega = \int_{\partial\Omega_F} \mathbf{t}_{\text{fluid}} \cdot \zeta \mathrm{d}\gamma - \int_\Omega \mathbb{T} : \nabla\zeta \mathrm{d}\Omega + \int_\Omega \rho_s \mathbf{b} \cdot \zeta \mathrm{d}\Omega. \tag{7.48}$$

For a linear elastic solid, the constitutive relation (7.42) is injected into equation (7.48):

$$\int_\Omega \mathbb{T} : \nabla\zeta \mathrm{d}\Omega = \int_\Omega \underbrace{\lambda (\nabla \cdot \mathbf{d})\mathbb{I} : \nabla\zeta \, \mathrm{d}\Omega}_{\nabla \cdot \mathbf{d}\nabla \cdot \zeta} + \int_\Omega \underbrace{2\mu\, \mathrm{e}(\mathbf{d}) : \nabla\zeta \, \mathrm{d}\Omega}_{\mathrm{e}(\mathbf{d}):\mathrm{e}(\zeta)}, \tag{7.49}$$

where in the second term we have used that $\mathrm{e} : \nabla\zeta = \mathrm{e} : (\nabla\zeta)^T$ since the small strain tensor is symmetric, and we have defined the small strain of the test function as $\mathrm{e}(\zeta) = (\nabla\zeta + (\nabla\zeta)^T)/2$. The resulting governing equation to be solved numerically is

$$\int_\Omega \rho_s \frac{\partial^2 \mathbf{d}}{\partial t^2} \cdot \zeta \mathrm{d}\Omega = \underbrace{-\int_\Omega \lambda\nabla \cdot \mathbf{d}\nabla \cdot \zeta \mathrm{d}\Omega - \int_\Omega 2\mu\mathrm{e}(\mathbf{d}) : \mathrm{e}(\zeta)\mathrm{d}\Omega}_{a(\mathbf{d},\zeta)}$$
$$\underbrace{+ \int_{\partial\Omega_f} \mathbf{t}_{\text{fluid}} \cdot \zeta \mathrm{d}\gamma + \int_\Omega \rho_s \mathbf{b} \cdot \zeta \mathrm{d}\Omega}_{B(\zeta)}, \qquad \forall \zeta \in V, \tag{7.50}$$

where $a(\mathbf{d}, \zeta)$ is the bilinear form of the problem, which depends both on the unknown field and the test function, while $B(\zeta)$ is a forcing term depending on the test function and the external hydrodynamic loads. V is the space of the solution and test functions, such as $H^1(\Omega)_{\partial\Omega_D}$, a subset of the Sobolev space satisfying the homogeneous Dirichlet condition at Ω_D.

The Galerkin method for the numerical approximation of problem (7.50) consists of finding an approximate solution $\mathbf{d}_h$, where h is the discretization parameter which indicates how small the discretization size is (in particular, if $h \to 0$, the discretization will have infinitely many degrees of freedom). We then identify a subspace $V_h \subset V$ of dimension $N < \infty$, with the formal requirement that $V_h \to V$ as $h \to 0$. The space discretized version of equation (7.50) can then be written as

$$\int_\Omega \rho_s \frac{\partial^2 \mathbf{d}_h}{\partial t^2} \cdot \zeta_h d\Omega = a(\mathbf{d}_h, \zeta_h) + B(\zeta_h). \tag{7.51}$$

Given a basis $\langle \varphi_1, \varphi_2, \ldots, \varphi_N \rangle$, the projection of the unknown components of the field $\mathbf{d}_h(\mathbf{x}, t)$ can be represented as

$$\mathbf{d}_h(\mathbf{x}, t) = \sum_{j=1}^{N} \kappa_j(t) \varphi_j(\mathbf{x}), \tag{7.52}$$

where the temporal functions of the modal coefficients of the (vectorial) displacement field, $\kappa_j(t)$, become the unknowns of the problem. Injecting equation (7.52) into the weak form of the momentum equation (7.51) one gets

$$\int_\Omega \rho_s \sum_{j=1}^{N} \frac{d^2\kappa_j(t)}{dt^2} \varphi_j(\mathbf{x}) \cdot \zeta_h d\Omega = a\left(\sum_{j=1}^{N} \kappa_j(t)\varphi_j, \zeta_h\right) + B(\zeta_h). \tag{7.53}$$

As any spatial component of the function ζ_h can be written as a linear combination of the basis functions φ_j, equation (7.53) reduces to a linear system of $N \times D$ equations with D the spatial dimensionality of the problem:

$$\mathbb{M}\frac{d^2 K}{dt^2} = \mathbb{A}K + B, \tag{7.54}$$

where K is a time varying array containing the $\kappa_j(t)$ with $j = 1, N$, and $\mathbb{A}$ is the stiffness matrix (Fish and Belytschko, 2007). The mass matrix $\mathbb{M}$ is block diagonal, having the matrix $m_{i,j} = \int_\Omega \rho_s \varphi_j(\mathbf{x})\varphi_i(\mathbf{x})d\Omega$ on its diagonal block for each field component. The array B depends on the instantaneous volume forcing term and the traction condition on the wet boundary. Equation (7.54) is in semidiscrete form as we focus here on the spatial discretization.

In the FEM, the structural domain is divided into a variety of subdomains (grids or meshes, which are continuous and disjoint) and on each subdomain

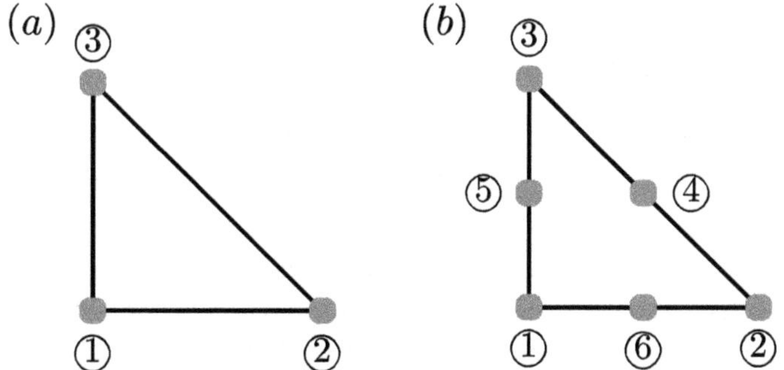

Figure 7.5 Reference element with (a) linear and (b) quadratic control points.

a certain number of points (or degrees of freedom) is distributed. In one-dimensional problems, linear segments are used, whereas two-dimensional domains are partitioned through triangles or quadrilaterals. From now on, we focus on the case of triangular partitions τ_h in two dimensions ($D = 2$), which allow for greater flexibility in handling complex shapes. As discussed in Chapter 3, a typical output for a mesh generator is given by the number of elements N_e, the number of vertices N_v, the number of boundary elements N_b, a list of vertices with their coordinates, and a list of triangles with the corresponding indices of their three vertices.

The finite-element space is defined as the set of all piecewise polynomials of degree r:

$$\mathcal{P}_r = \left\{ \varphi(x, y) = \sum_{i+j \leq r} a_{ij} x^i y^j \right\}, \tag{7.55}$$

where the number of polynomial terms (and of the degrees of freedom) over each triangle $\mathcal{T}$ is $\dim(\mathcal{P}_r) = (r + 1)(r + 2)/2$. As an example for linear and quadratic polynomials we have:

$$\begin{aligned}
\mathcal{P}_1 &= \{\varphi(x, y) = a_{0,0} + a_{1,0}x + a_{0,1}y\}, \\
\mathcal{P}_2 &= \left\{\varphi(x, y) = a_{0,0} + a_{1,0}x + a_{0,1}y + a_{2,0}x^2 + a_{0,2}y^2 + a_{1,1}xy\right\}.
\end{aligned} \tag{7.56}$$

The local polynomials (7.56) are defined over each subdomain of the partition, simply streched to fit the shape and orientation of any triangle $\mathcal{T}$ by adjusting the coefficients $a_{i,j} \in \mathbb{R}$. Therefore, it is practical to define the FEM basis on a reference element, which is then mapped through a linear transformation to any local element $\mathcal{T}$. For triangular grids, the reference element is generally the unitary rectangular triangle lying on the plane (ξ, η); see Figure 7.5a. For linear elements ($r = 1$), the local nodal functions are

$$\varphi_1 = 1 - \xi - \eta, \qquad \varphi_2 = \xi, \qquad \varphi_3 = \eta, \tag{7.57}$$

which are equal to 1 on the ith node and 0 on the other nodes. On the other hand, for quadratic elements ($r = 2$; see Figure 7.5b) the local functions are

$$\varphi_1 = (1 - \xi - \eta)(1 - 2\xi - 2\eta), \qquad \varphi_2 = \xi(2\xi - 1), \qquad \varphi_3 = \eta(2\eta - 1),$$
$$\varphi_4 = 4\xi\eta, \qquad \varphi_5 = 4\eta(1 - \xi - \eta), \qquad \varphi_6 = 4\xi(1 - \xi - \eta),$$

$$(7.58)$$

which take unit value on the corresponding node and vanish in all of the other five nodes. The next step is the transformation of coordinates that map the local functions of the reference element into those of the kth triangle element. The latter is defined by the vertices A, B and C, hence the vector that connects the origin to any point P of the triangle is given by

$$\mathbf{v}_{OP} = \mathbf{v}_{OA} + \xi\mathbf{v}_{AB} + \eta\mathbf{v}_{AC}, \qquad (7.59)$$

where with $\mathbf{v}_{OP}$ and $\mathbf{v}_{OA}$ we indicate the vector positions of the points P and A with respect to the origin O, whereas $\mathbf{v}_{AB}$ ($\mathbf{v}_{AC}$) is the vector corresponding to the triangle edges between the points A and B (C). Componentwise, equation (7.59) reads

$$\begin{pmatrix} x \\ y \end{pmatrix} = \underbrace{\begin{bmatrix} x_B - x_A & x_C - x_A \\ y_B - y_A & y_C - y_A \end{bmatrix}}_{\mathbb{B}} \begin{pmatrix} \xi \\ \eta \end{pmatrix} + \begin{pmatrix} x_A \\ y_A \end{pmatrix}, \qquad (7.60)$$

where the matrix $\mathbb{B}$ of the linear transformation (7.60) corresponds to the Jacobian matrix. Then the integrals over any triangle $\mathcal{T}$ belonging to the partition of the domain can be evaluated in the reference element, for example:

$$\int_{\mathcal{T}} \varphi_j(x, y) \cdot \varphi_i(x, y)\mathrm{d}x\mathrm{d}y = \int_0^1 \int_0^{1-\eta} \varphi_j(\xi, \eta) \cdot \varphi_i(\xi, \eta)\,\det(\mathbb{B})\mathrm{d}\xi\mathrm{d}\eta. \qquad (7.61)$$

In a similar fashion, all the entries of the mass and stiffness matrices in equation (7.54) can be computed.

7.4.2 The Method of Interaction Potentials

A possible emerging alternative for the modeling of deformable body dynamics is based on the interaction potentials approach (Fedosov et al., 2010), in which the continuum is discretized by triangles or tetrahedra and the mass is distributed among the element nodes that are connected by springs to form an elastic network (Figure 7.6). Such a spring-network approach yields discrete models, defining the elastic forces on the vertices based on edge lengths and reciprocal angle variations of the triangles. Also in this case, we briefly introduce the method for two-dimensional thin structures, although it can be extended to three-dimensional continua (Viola et al., 2020, 2023a).

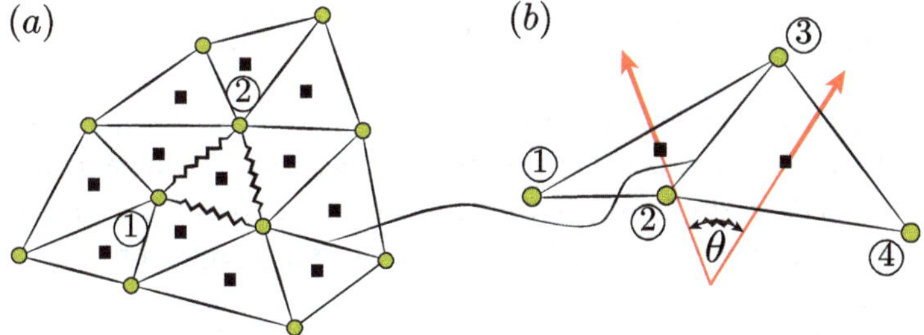

Figure 7.6 Sketch of the structural solver based on the interaction potentials method. The medium is discretized using a triangular mesh and (a) a spring is placed at each edge, thus providing the in-plane stiffness of the material. (b) The bending stiffness of the shell is then obtained by the out-of-plane springs connecting the centroids of two adjacent triangular faces sharing an edge.

The mass of the structure is lumped at the vertices of the triangles proportionally to the areas of the triangles sharing a given node. The mass of a triangular face with surface A_{fj} and density ρ_{fj} is equally distributed among its three nodes, and the mass of a node, m_n is equal to

$$m_n = \frac{1}{3} \sum_{j=1}^{N_{nf}} \rho_{fj} s_{fj} A_{fj}, \tag{7.62}$$

with the summation extended only to the N_{nf} triangles sharing the selected node n. The material density ρ_{fj} can be non-uniform over the surface, as well as the local thickness s_{fj}.

As shown in Figure 7.6a, adjacent nodes sharing an edge are connected by elastic springs oriented as the mesh edges, generating an internal stress τ when deforming. Hence, given an edge of length l connecting the nodes $n1$ and $n2$ with local thickness s, the stress τ should be translated into the corresponding force applied at its tips:

$$\mathbf{F}_{n1}^{el} = \tau \underbrace{s\frac{A_1 + A_2}{l}}_{\text{structure cross-section}} \underbrace{\frac{\mathbf{r}_{n1} - \mathbf{r}_{n2}}{l}}_{\text{force direction}}, \qquad \mathbf{F}_{n2}^{el} = -\mathbf{F}_{n1}^{el}, \tag{7.63}$$

with $\mathbf{r}_{n1}$ ($\mathbf{r}_{n2}$) being the position of the node $n1$ ($n2$) and $A_{1,2}$ the area of the triangles sharing the edge. Note that the first underbraced term multiplying the stress corresponds to the structure cross section with respect to the edge, while the second one is the unitary vector of the elastic force direction. For a linear elastic material, the constitutive relation $\tau = E\epsilon$ holds, where E is the Young modulus and ϵ is the local stretching calculated as $\epsilon = (l - l_0)/l$, with l and l_0 the actual and the stress-free length of the edge, respectively. Hence, equation (7.63) can be rewritten as Hooke's force:

$$\mathbf{F}_{n1}^{el} = k_e(l - l_0)\frac{\mathbf{r}_{n1} - \mathbf{r}_{n2}}{l}, \quad \text{with } k_e = \frac{Es(A_1 + A_2)}{l^2}, \tag{7.64}$$

which is associated with the elastic potential $W^{el} = \frac{1}{2}k_e(l - l_0)^2$. The same approach can also be extended to hyperelastic materials (Hammer et al., 2011; de Tullio and Pascazio, 2016), as in the case of a Fung-type constitutive relation, by adopting the corresponding hyperelastic stress τ in equation (7.63) (Viola et al., 2023a).

Since in the two-dimensional spring network the axial loading (7.63) accounts only for the in-plane stiffness, in the case of shells or plates, an additional bending energy term has to be included to provide the out-of-plane stiffness. The out-of-plane deformation of two adjacent triangles sharing an edge is then associated with an elastic energy due to the contraction/expansion of a bending spring, whose energy involves four adjacent nodes, as shown in Figure 7.6b. Considering a surface with nonzero reference curvature in the stress-free configuration, the discretized bending energy can be written as (Kantor and Nelson, 1987)

$$W_b = k_b[1 - \cos(\theta - \theta_0)], \tag{7.65}$$

where θ is the angle between the normals of adjacent triangular faces of the tessellated surface, and θ_0 is the equilibrium angle determined by the spontaneous curvature, that is the shell curvature at the initial stress-free state. The bending constant, k_b, is related to the bending modulus of the structure, B, according to $k_b = 2B/\sqrt{3}$, with $B = Es^3/[12(1 - v_m^2)]$ for a planar structure, where v_m is the Poisson ratio of the material. The corresponding bending nodal forces, $\mathbf{F}_n^{be}$, can then be obtained by taking the gradient of the potential (Li et al., 2005; de Tullio and Pascazio, 2016):

$$\begin{aligned}
\mathbf{F}_{n1}^{be} &= \beta_b[b_{11}(\mathbf{n}_1 \times \mathbf{r}_{32}) + b_{12}(\mathbf{n}_2 \times \mathbf{r}_{32})], \\
\mathbf{F}_{n2}^{be} &= \beta_b[b_{11}(\mathbf{n}_1 \times \mathbf{r}_{13}) + b_{12}(\mathbf{n}_1 \times \mathbf{r}_{34} + \mathbf{n}_2 \times \mathbf{r}_{13}) + b_{22}(\mathbf{n}_2 \times \mathbf{r}_{34})], \\
\mathbf{F}_{n3}^{be} &= \beta_b[b_{11}(\mathbf{n}_1 \times \mathbf{r}_{21}) + b_{12}(\mathbf{n}_1 \times \mathbf{r}_{42} + \mathbf{n}_2 \times \mathbf{r}_{21}) + b_{22}(\mathbf{n}_2 \times \mathbf{r}_{42})], \\
\mathbf{F}_{n4}^{be} &= \beta_b[b_{12}(\mathbf{n}_1 \times \mathbf{r}_{23}) + b_{22}(\mathbf{n}_2 \times \mathbf{r}_{23})],
\end{aligned} \tag{7.66}$$

with

$$b_{11} = -\frac{\cos\theta}{|\mathbf{n}_1|^2}; \qquad b_{12} = \frac{1}{|\mathbf{n}_1||\mathbf{n}_2|}; \qquad b_{22} = -\frac{\cos\theta}{|\mathbf{n}_2|^2} \tag{7.67}$$

and

$$\beta_b = k_b\frac{\sin\theta\cos\theta_0 - \cos\theta\sin\theta_0}{\sqrt{1 - \cos^2\theta}}. \tag{7.68}$$

Additional energy terms can be added in order to enforce further constraints on the area (either local or global) and on the volume (if the surface is closed) (de Tullio and Pascazio, 2016).

For each nth node, the equation of motion governing its dynamics is

$$m_n \ddot{\mathbf{x}}_n = \mathbf{F}_n^{ext} + \mathbf{F}_n^{int}, \tag{7.69}$$

where m_n is its mass, $\ddot{\mathbf{x}}_n$ is the acceleration, $\mathbf{F}_n^{ext}$ are all external forces (i.e. hydrodynamic loads, volume forces) and $\mathbf{F}_n^{int} = \mathbf{F}_n^{el} + \mathbf{F}_n^{be}$ are the internal forces.

If needed, an additional damping term could be added on the right-hand side in order to take into account the viscoelasticity of the structure or other dissipative phenomena.

This method, commonly used in computer graphics (Nealen et al., 2006), is very advantageous in the current context owing to its reduced computational cost compared to traditional FEMs. However, it must be pointed out that it does not represent a true continuum approach, as it models the structure using a discrete network of concentrated masses linked by elastic elements. In this mesoscale framework, the structure continuum features emerge only at scales larger than the individual elements. Additionally, it is crucial to avoid artificial anisotropy effects produced by excessively regular discretization with the element edges or faces aligned along preferential directions (Van Gelder, 1998).

<h1 style="text-align:center">8</h1>

Turbulence and Wall Models within IBMs

As already mentioned, CFD simulations become computationally more demanding as the Reynolds number increases, owing to the decreasing size attained by the smallest flow structures. In fact, an eddy at a scale ℓ entails grid spacing of comparable dimension (actually, the grid spacing should be half the size of the eddy being described, according to the Nyquist-Shannon sampling theorem, as discused in section 2.3) in order to avoid its discrete misrepresentation.

Indeed, the theory of turbulence predicts that a body of size L in a flow at Reynolds number Re can produce, in the bulk flow, scales as small as $\eta \approx LRe^{-0.75}$, while next to the solid boundaries $\delta^+ \approx LRe^{-0.90}$: The former is referred to as the Kolmogorov scale and the latter as the viscous length or wall unit.

Since δ^+ decreases with Re faster than η, the fine resolution of the wall regions is not necessary in the bulk and computational resources are usually saved by using non-uniform discretizations in the wall-normal direction.

In typical applications, the Reynolds number is large enough for δ^+ to yield the most restrictive limitation with the amount of mesh elements within the boundary layers that easily exceeds 90 percent of the total (Piomelli, 2008). Considering a configuration like that of Figure 8.1a, the dynamics of the thin boundary layers developing at the body surface is described by a mesh proportional to δ^+ and uniform, in the wall-parallel directions, while the discretization can benefit from a non-uniform grid spacing proportional to the wall distance, in the wall-normal direction. Following Pope (2000), using the correlation between L/δ^+ and Re, the total number of grid points per cubic L scales like $N_{BF} \sim Re^{1.8} \log Re$ (this result is obtained by assuming a uniform discretization in the wall parallel directions along with a grid spacing proportional to the wall distance in the normal direction).

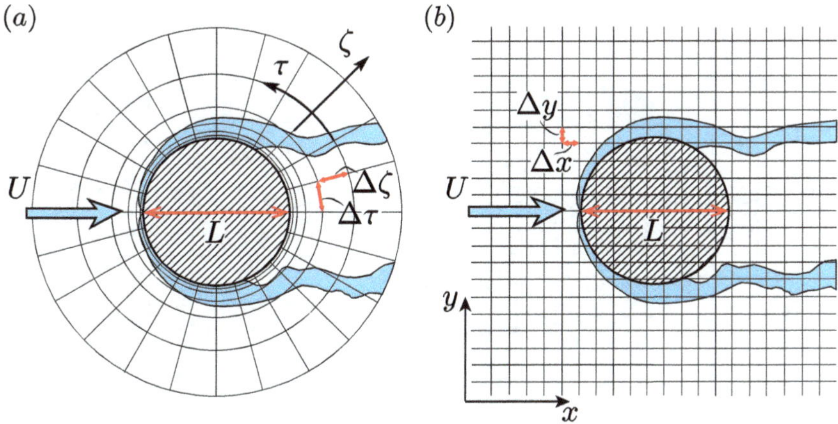

Figure 8.1 Sketches of (a) body-fitted and (b) non-body-conformal discretizations for the flow around a solid body.

The context of IBMs is that of Figure 8.1b, where the mesh is non-body-conformal and the absence of wall-normal and wall-parallel isolines precludes the possibility of clustering the mesh within near-wall layers.

Depending on the specific surface point, any direction can be normal or tangential; thus, the distribution of the nodes cannot benefit from anisotropic clustering and it must be proportional to δ^+ in all directions. This implies that the total number of nodes per cubic L now scales as $N_{IB} \sim Re^{2.7}$, which grows with Re at a faster rate than N_{BF}.

It is worth stressing that the time to solution of a CFD simulation depends on the mesh size and also on the numerical algorithm; therefore, even if for a given Re the result is always $N_{IB} > N_{BF}$, IBM simulations can still be faster than the body-fitted counterpart because of the simpler and more efficient solution algorithm implemented on regular, structured and orthogonal meshes. Nevertheless, the estimates for the grids in the two cases yield the ratio $N_{IB}/N_{BF} \sim Re^{0.9}/\log Re$, which eventually makes the IB approach infeasible for high-Re flows.

Just as for standard CFD simulations, the computational overhead originates from the near-wall region, and classical remedies to alleviate the problem are wall models (Piomelli, 2008) or adaptive grid refinement (Prouvost et al., 2024). In the context of immersed boundaries, additional care is necessary since these models must be combined with all the forcings, interpolations and reconstructions needed to enforce the boundary conditions on coordinate lines which do not comply with walls.

Figure 8.2 shows a cartoon of the two techniques applied to the same problem described in Figure 8.1; we will see that each method has advantages but also limitations, which depend on the specific set of equations used to model the flow and the way turbulence is modeled.

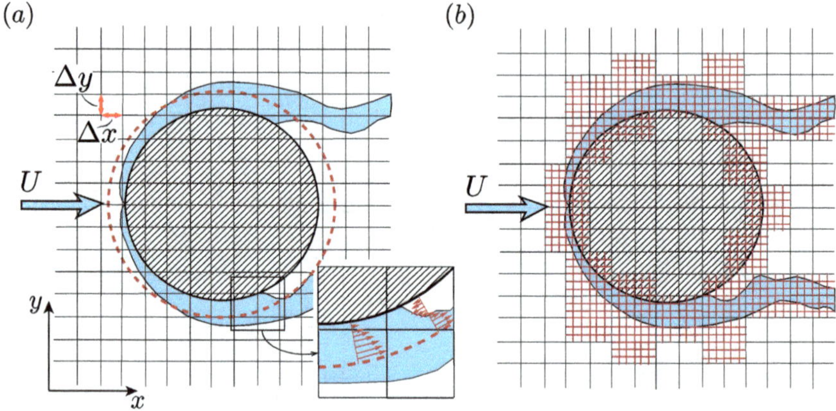

Figure 8.2 Sketches of (a) wall modeling and (b) adaptive mesh refinement for the flow around a solid body simulated on a non-body-conformal mesh by immersed boundary methods.

Turbulence modeling would deserve a volume on its own, and indeed several comprehensive textbooks are already available (Wilcox, 1998; Pope, 2000; Launder and Sandham, 2002; Durbin, 2021): Here we only introduce a few basic concepts in order to better explain the interaction of adaptive mesh refinement and wall models with immersed boundary methods when turbulence modeling is necessary.

8.1 Turbulent Flow Simulations

In Chapter 2 we derived the equations for mass conservation and balance of momentum that, for an incompressible, viscous and Newtonian fluid read

$$\nabla \cdot \mathbf{u} = 0, \tag{8.1}$$

$$\frac{\partial \mathbf{u}}{\partial t} + \nabla \cdot (\mathbf{uu}) = -\frac{\nabla p}{\rho} + \nu \nabla^2 \mathbf{u} + \mathbf{f}. \tag{8.2}$$

This is a closed system of four equations in the four unknowns p, $\mathbf{u} = (u, v, w)$ (density ρ is assumed constant) which, in principle, could be solved for any fluid viscosities ν to determine the unknown fields.

However, in Figure 2.4 and in the subsequent discussion, we have already shown that, for decreasing values of ν, the flow dynamics becomes richer both in time and space, making its numerical simulation more demanding. In Section 2.4 we have already anticipated that this is because of the energy cascade from large to small scales caused by nonlinear terms, while in Section 2.3 we have briefly discussed the consequences on the spatial numerical resolution and its implications. Here we want to point out that, paradoxically, the cause of these issues is the completeness of equations (8.1) and (8.2), which contain all the details of the flow, from the velocity gradient at the wall or the size of the recirculation bubble up to the smallest eddy in the wake that is about to be dissipated. This huge amount of information makes the aforementioned equations numerically intractable for most practical problems which need alternative formulations.

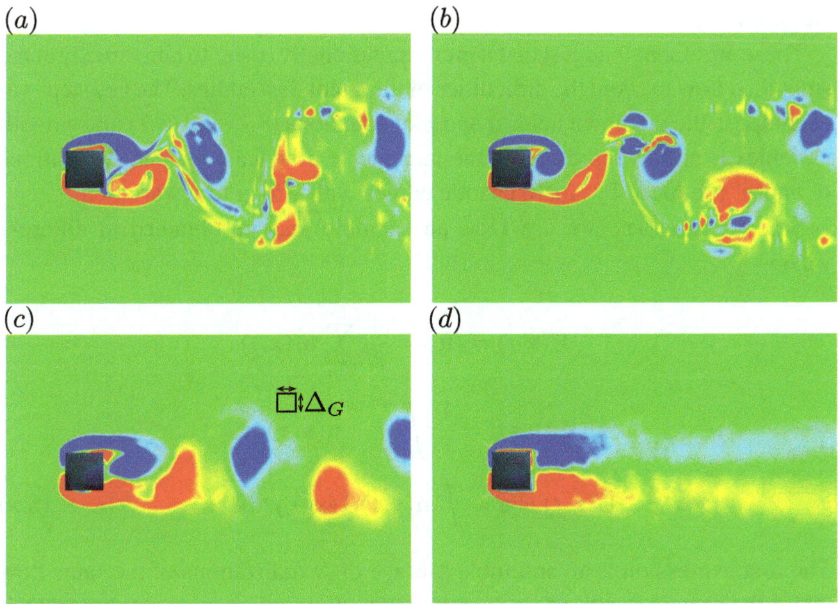

Figure 8.3 (a) and (b) Instantaneous snapshots of out-of-plane vorticity in the vertical symmetry plane for the three-dimensional flow around a square cylinder at $Re = 500$. (c) Instantaneous snapshot of the same flow as (a) and (b) but for the field filtered with a top-hat function of Δ_G support in the two dimensions of the section (the field has also been averaged in the out-of-plane homogeneous direction). (d) Time average of the field for $\approx 50D/U$ time units (D is the square edge and U the free-stream velocity).

These arguments are illustrated in Figures 8.3a,b, showing instantaneous snapshots of the flow around a square cylinder together with the field obtained by a smoothing low-pass filter defined later (Figure 8.3c) and that from the superposition of fields collected at different times (Figure 8.3d) which, for a statistically steady flow, is equivalent to making an ensemble average.

From panels (a) and (b) of Figure 8.3 we can appreciate the unsteady character of the flow and the richness in small scales that are described at the cost of a fine spatial discretization and a small time step; the flow, however, must be advanced for a long enough time to span the dynamics of the slow, large scales and to sample a sufficient number of statistically independent fields whose average yields converged statistics.

On the other hand, the large-scale and mean flows, in panels (c) and (d) of Figure 8.3, are characterized by smoother spatial structure and, comparing the near-wall region with that of the instantaneous configurations, it shows limited variation.

These arguments suggest that it is computationally easier to aim directly at the smoothed flow to avoid the difficulties of the small, fast eddies. The key step is to decompose the unknown fields $\mathbf{u}$ and p of equations (8.1) and (8.2) into smooth variables $\mathbf{U}$ and P, representing the large-scale flow features, and fluctuations $\mathbf{u}'$ and p' produced by small, chaotic eddies with strong time variation.

Smooth variables (we use $\mathbf{U}$ as an example) can be obtained in different ways:

$$\mathbf{U}(\mathbf{x}, t) = \langle \mathbf{u}_i \rangle = \frac{1}{N} \sum_{i=1}^{N} \mathbf{u}_i(\mathbf{x}, t),$$

$$\mathbf{U}(\mathbf{x}, t) = \overline{\mathbf{u}} = \frac{1}{T} \int_{t-T/2}^{t+T/2} \mathbf{u}(\mathbf{x}, \tau) d\tau,$$

$$\mathbf{U}(\mathbf{x}, t) = \widehat{\mathbf{u}} = \int \mathbf{u}(\mathbf{y}, t) G(\mathbf{x} - \mathbf{y}) d\mathbf{y}. \tag{8.3}$$

The first expression is an ensemble average of N realizations of the same flow while the second is a running average over a time window of amplitude T: If the flow is statistically steady, in the limit $N \to \infty$ and $T \to \infty$, these two procedures yield the same result which converges to a steady field $\mathbf{U}(\mathbf{x})$.

The third of equations (8.3) is a convolution of the field $\mathbf{u}$ with G that is a compactly supported function different from zero only in a region of size Δ_G around point $\mathbf{x}$; this procedure returns a filtered field $\widehat{\mathbf{u}}$ whose features smaller than Δ_G are smoothened out or completely eliminated, depending on the specific form of G.

Each of equations (8.3) can be thought of as an operator applied to a field (here, the velocity $\mathbf{u}$); therefore it could be applied to all terms of equations (8.1) and (8.2) to obtain the governing equations for the smooth fields.

Some standard manipulation of the original and averaged equations (see Pope (2000) and Mathieu and Scott (2000)) yields

$$\nabla \cdot \mathbf{U} = 0,$$

$$\frac{\partial \mathbf{U}}{\partial t} + \nabla \cdot (\mathbf{U}\mathbf{U}) - \nabla \cdot \mathbf{R}_T = -\frac{\nabla P}{\rho} + \nu \nabla^2 \mathbf{U} + \mathbf{F}, \tag{8.4}$$

which is almost identical to the initial system of equations, except for the extra quantity $\nabla \cdot \mathbf{R}_T$ coming from the nonlinear term. P and $\mathbf{F}$ are the filtered pressure and forcing terms. In fact, considering as an example the first of equations (8.3) applied to the convective term of equation (8.2), we have:

$$\langle \nabla \cdot (\mathbf{u}\mathbf{u}) \rangle = \nabla \cdot \langle (\mathbf{U} + \mathbf{u}')(\mathbf{U} + \mathbf{u}') \rangle = \nabla \cdot \langle \mathbf{U}\mathbf{U} \rangle + \nabla \cdot \langle \mathbf{U}\mathbf{u}' \rangle$$

$$+ \nabla \cdot \langle \mathbf{u}'\mathbf{U} \rangle + \nabla \cdot \langle \mathbf{u}'\mathbf{u}' \rangle = \nabla \cdot \mathbf{U}\mathbf{U} + \nabla \cdot \langle \mathbf{u}'\mathbf{u}' \rangle, \tag{8.5}$$

in which we have used the results $\langle \mathbf{U} \rangle = \mathbf{U}$ and $\langle \mathbf{u}' \rangle = \mathbf{0}$ and the fact that even if the latter is true, it is $\langle \mathbf{u}'\mathbf{u}' \rangle \neq \mathbf{0}$.

Note that now $\mathbf{R}_T \equiv -\langle \mathbf{u}'\mathbf{u}' \rangle$ (or $\equiv -\overline{\mathbf{u}'\mathbf{u}'}$ or $\equiv -\widehat{\mathbf{u}'\mathbf{u}'}$ if the second or the third of (8.3) is used[1]) is an unknown symmetric tensor which makes the system (8.4) not closed since the number of available equations does not match that of the unknowns.

This is known as the *turbulence closure* problem and it entails modeling assumptions, as any attempt to derive an exact equation for $\mathbf{R}_T$ from those for $\mathbf{u}$ and $\mathbf{U}$ produces higher-order unknowns, like third-order correlation of $\mathbf{u}'$ and its correlation with pressure fluctuations p', which widens the gap between equations and unknowns (Tennekes and Lumley, 1972).

It can be shown that $\mathbf{R}_T$ acts on the mean flow as an additional stress that takes energy from the large-scale flow to feed the turbulent fluctuations; for this reason it is referred to as the turbulent-stress or Reynolds-stress tensor. Using the same arguments as for the constitutive relations of Newtonian fluids, the deviatoric part of $\mathbf{R}_T$ stresses can be obtained from the rate of strain of the smooth flow $\mathbf{S} = (\nabla \mathbf{U} + (\nabla \mathbf{U})^T)/2$ according to

$$\mathbf{R}_T - \frac{1}{3} Tr(\mathbf{R}_T)\mathbf{I} = \boldsymbol{\tau}_T = 2\nu_T \mathbf{S}, \tag{8.6}$$

where $Tr(\cdot)$ is the trace of the tensor and ν_T is the unknown field of turbulent viscosity which depends on the local flow features. The aforementioned

---

[1] Indeed, if the fields are filtered, like in the third of equations (8.3), the details are more complex since, not only $\widehat{\mathbf{U}} \neq \mathbf{U}$ and $\widehat{\mathbf{u}'} \neq \mathbf{0}$, but the filtering operator $\widehat{\ \ }$ does not commute with the spatial differential operators of (8.2), except for special functions G. The textbook by Pope (2000) discusses in a comprehensive way the details; here we will not indulge further on the matter, on account of the possibility of including everything into $\mathbf{R}_T$.

assumption is referred to as the Boussinesq hypothesis and, at first glance, it appears just to move an unknown ($\mathbf{R}_T$ or τ_T) into another (ν_T). However, τ_T is a symmetric tensor with six independent unknown fields, while ν_T is a single scalar field, with the result that equation (8.6) reduces the unknowns from six to one without physical justification. An important consequence of this closure is that the eigenvectors of τ_T are always locally and instantaneously aligned with those of $\mathbf{S}$; this can be possible only if the smooth flow has regular enough dynamics to allow the two tensors to equilibrate.

Despite all issues, the vast majority of turbulence models rely on the Boussinesq hypothesis in the form of equation (8.6), even if equations (8.4) can be resolved at very disparate scales and the required modeling of τ_T entails different physics.

8.2 RANS Modeling

The oldest and most common approach is to employ the second of equations (8.3) to extract only the mean (or slow-varying) fields and model any fluctuations as turbulence; the advantage is that $\mathbf{U}$ and P are very smooth and their simulation requires limited computational effort. On the other hand, the fluctuations $\mathbf{u}'$ and p' are likely to contain not only random turbulent motion but also deterministic, unsteady flow features whose modeling is extremely problem-dependent. This technique is known as RANS (Reynolds-averaged Navier–Stokes) and is very popular among industrial applications characterized by high Reynolds numbers and complex geometrical configurations.

Equations (8.4) with (8.6) become

$$\nabla \cdot \mathbf{U} = 0,$$

$$\frac{\partial \mathbf{U}}{\partial t} + \nabla \cdot (\mathbf{UU}) = -\frac{\nabla P^*}{\rho} + 2\nabla \cdot [(\nu + \nu_T)\mathbf{S}] + \mathbf{F}, \qquad (8.7)$$

whose solution is possible, provided the eddy viscosity is defined. Note that $P^* = P - \rho Tr(\mathbf{R}_T)/3$ is a modified pressure which coincides with the pressure P at the wall.

There is plenty of literature devoted to different definitions of ν_T although all models are derived on the basis of dimensional arguments, consistency with experimental observation and physical arguments, or practical aspects such as simplicity and generality. The various models are classified in terms of the number of equations to be solved, in addition to equations (8.7), to determine ν_T.

There are (i) zero-equation or algebraic models, such as the mixing length model or the Baldwin–Lomax model; (ii) one-equation models, such as the

Spalart–Allmaras model; (iii) two-equation models, such as the k–ϵ and the k–ω models; and (iv) the anisotropic or Reynolds-stress models, entailing a tensor equation for $\mathbf{R}_T$, which are the only ones not relying on the Boussinesq hypothesis.

The classic textbook Wilcox (1998) is an excellent reference for the various descriptions and implementation details of all models; here we mention only that in the context of IBMs, any additional quantities or equations introduced to determine ν_T must be provided with boundary conditions and reconstruction procedures to impose values at the immersed interface.

Among many possibilities, we focus our attention on a two-equation hybrid model k–ω/k–g (more often, simply referred to as k–g) (Kalitzin, 1997) since it has been customized for numerical methods with IBM procedures (Kalitzin and Iaccarino, 2002; Kalitzin et al., 2005).

The base of the model is the k–ω, as described in Wilcox (1998), which defines a turbulent viscosity as $\nu_T = k/\omega$ on dimensional arguments from the turbulent kinetic energy per unit mass $k = \langle \mathbf{u}' \cdot \mathbf{u}' \rangle/2$ and the specific dissipations rate $\omega = \epsilon/k$ (ϵ is the turbulent energy dissipation rate).

Although exact equations for k and ω could be derived from equations (8.2) and (8.7), these would incur the same closure problems as the equation for $\mathbf{R}_T$, therefore they are assembled as a model themselves:

$$\frac{\partial k}{\partial t} + \mathbf{U} \cdot \nabla k = P_k - c_\mu \omega k + \nabla \cdot [(\nu + \sigma_k \nu_T)\nabla k], \tag{8.8}$$

$$\frac{\partial \omega}{\partial t} + \mathbf{U} \cdot \nabla \omega = \frac{\gamma_1 \omega P_k}{k} - \beta_1 \omega^2 + \nabla \cdot [(\nu + \sigma_\omega \nu_T)\nabla \omega], \tag{8.9}$$

where

$$P_k = \nu_T S^2, \qquad S = \sqrt{2\mathbf{S} : \mathbf{S}}, \tag{8.10}$$

with the five model constants which, for the original model, are:

$$\sigma_k = \sigma_\omega = 0.5; \quad \gamma_1 = 5/9; \quad \beta_1 = 0.075; \quad \text{and} \quad c_\mu = 0.09. \tag{8.11}$$

It is worth mentioning that these values do not come from theoretical considerations but rather are tuned to reproduce, with the model, benchmark results. Furthermore, $\mathbf{R}_T$ in RANS contains everything but the mean flow and its problem-dependent features entail specific changes to the constants.

The solution of equations (8.8) and (8.9) requires the imposition of boundary conditions for the variables k and ω: The former is straightforward as, regardless of the Reynolds number, next to a solid boundary a laminar viscous sublayer forms where $\mathbf{u}' = \mathbf{0}$ and, therefore, $k = 0$.

On the other hand, approaching a no-slip boundary ϵ grows to large, finite values; thus the specific dissipation rate ω tends to infinity and, more specifically, it results in $\omega \sim 1/\zeta^2$, where ζ is the wall-normal distance.

Clearly, in any numerical method it is impossible to impose such a boundary condition, and Wilcox (1998) describes the errors associated with the numerical integration of ω up to the wall. As a remedy, Menter (1993) suggested using the approximate condition:

$$\omega = \frac{60\nu}{\beta_1 \zeta_1^2},\tag{8.12}$$

where ζ_1 is the distance between the center of the first cell and the wall.

The fact that ω exhibits the largest gradients near the wall and diverges at the boundary poses a serious obstacle to the forcing and reconstruction steps of IBMs which, instead, are based on linear or polynomial interpolations.

In order to avoid these issues, Kalitzin and Iaccarino (2002) proposed a simple and ingenious change of variable $g = 1/\sqrt{c_\mu \omega}$ that eliminates all the above limitations. In fact, on account of equation (8.12) and the variable definition, it results $g \sim \zeta$ in the near-wall region which complies perfectly with a linear IB reconstruction and yields the trivial wall boundary condition $g = 0$.

It is worth noticing also that the popular two-equation k–ϵ RANS model is affected by a similar problem since the kinetic energy dissipation rate per unit mass ϵ has the peak value at the wall and steeply decreases with the wall distance ($\sim \zeta^{-2}$) thus preventing the standard IB reconstructions from properly representing its behavior.

Equations (8.8) and (8.9) recast for the couple k–g read

$$\frac{\partial k}{\partial t} + \mathbf{U} \cdot \nabla k = P_k - \frac{k}{g^2} + \nabla \cdot [(\nu + \sigma_k \nu_T)\nabla k],\tag{8.13}$$

$$\frac{\partial g}{\partial t} + \mathbf{U} \cdot \nabla g = \frac{-\gamma_1 g P_k}{2} + \frac{\beta_1}{2gc_\mu} - (\nu + \sigma_g \nu_T)\frac{3}{g}\nabla g \cdot \nabla g + \nabla \cdot [(\nu + \sigma_g \nu_T)\nabla g],\tag{8.14}$$

with $\sigma_g = 0.5$ and the turbulent viscosity defined as $\nu_T = c_\mu k g^2$.

The k–g model has been designed to avoid the divergence of ω at the wall although it has to cope with the dual problem far from solid boundaries, where turbulence intensity is typically low, $\omega \to 0$ and $g \to \infty$, which makes infeasible the use of the model for flows with laminar or low-dissipation regions. For this reason Kalitzin and Iaccarino (2002) have proposed a hybrid implementation using the k–g variables close to the walls and the standard k–ω variables in the remaining domain: Several validation tests have shown that indeed this strategy allows RANS simulations of flows with high Reynolds numbers, provided adequate grid resolution could be assured in the near-wall region, which can seldom be achieved with immersed boundary techniques without body-conformal meshes. This issue will be discussed in Section 8.4 when introducing wall models.

8.3 LES Modeling

A relatively more recent technique to tackle the governing equations for high Reynolds numbers is based on the third of (8.3); this technique filters the equations at a fixed cutoff length Δ_G and models as turbulent fluctuations only the eddies smaller than that scale. Usually, Δ_G is identified with the local mesh size Δ (implicit filtering), although there are more sophisticated approaches (Germano et al., 1991; Germano, 1992) based on multiple cutoff scales that require explicit filtering. If Δ_G is fine enough to be in the middle of the inertial range of turbulence spectrum, according to the Kolmogorov hypotheses (Batchelor, 1982), the fluctuations $\mathbf{u}'$ and p' tend to have dynamics independent of the specific flow and they can be parametrized by reliable, universal turbulence models. This approach is known as LES (large-eddy simulation) with the meaning that equations (8.4) describe directly the dynamics of the large eddies ($> \Delta_G$) while the effect of the smaller structures, referred to as subgrid scales, on the mean flow is accounted for by the eddy viscosity ν_T.

Relying on a "mixing length" assumption, this eddy viscosity is computed as the product between a length ℓ and a velocity $\mathcal{V}$: The former is taken as the filter width $\ell = \Delta_G = \Delta$ and the latter is estimated as $\mathcal{V} \approx \ell S$ (where S is defined as in (8.10)). The combination of these assumptions yields

$$\nu_T = (C_\nu \Delta)^2 S, \tag{8.15}$$

with C_ν a parameter, the Smagorinsky constant (Smagorinsky, 1963), to be determined. The most common assumption of LES models is that locally and instantaneously turbulent energy production and dissipation are in equilibrium: Using the spectrum of homogeneous isotropic turbulence, it can be shown (Pope, 2000) that this condition yields $C_\nu \approx 0.17$ for almost every filter function, and this value is consistent with the range $0.15 \leq C_\nu \leq 0.2$ commonly adopted in simulations.

When compared to RANS, an evident advantage of LES modeling is that it needs only a single external constant C_ν which, moreover, shows very limited dependence on the specific problem.

Even this constant can be determined as part of the solution following the procedures described in Germano et al. (1991) and Lilly (1992); they apply an implicit filter and an explicit filter to compute the same tensor fields at two different scales (Δ_G and $2\Delta_G$) and they must obey the identity of Germano

(1992) from which the "constant" is computed. Since in this case C_v depends, in principle, on space and time, the resulting model is called *dynamic*.

Another advantage of LES with respect to RANS is that the v_T computation from equation (8.15) needs only S, which comes from the filtered $\mathbf{U}$ of (8.4) without auxiliary fields in turn obtained by model equations. The simplicity of the formulation, however, is largely made up for by the refined meshes needed to maintain the filter width well within the inertial range of the turbulent spectrum, and this makes LES more computationally demanding than RANS, which is the reason for the limited application of the former technique as a routine industrial tool.

Concerning the combination of LES with IBMs, the absence of additional equations makes the implementation easier than RANS. The only delicate point is the treatment of eddy viscosity which, close to solid boundaries, behaves as $v_T \sim \zeta^3$. Although this scaling could be handled by an ad hoc reconstruction, using the auxiliary variable $\chi = \sqrt[3]{v_T}$ allows the standard linear reconstruction used for momentum to be applied also for eddy viscosity.

Among the first results obtained with this strategy are those by Verzicco et al. (2000, 2004), whose flows had a small enough Reynolds number so that sufficient wall resolution was achieved even without a body-fitted mesh. However, this is not the case for the high Reynolds number characterizing most applications, and additional wall modeling is necessary.

8.4 Wall Models

A relevant observation is in order for the flow in the near-wall region: Looking at the panels of Figure 8.3 it is evident that while filtering or averaging the flow (Figure 8.3c,d) considerably smooth the bulk with respect to instantaneous snapshots (Figure 8.3a,b), the same is not true for the shear layers next to solid boundaries which show basically the same steep gradients in all panels. The undesired consequence is that the mesh resolution at the wall needed to capture the boundary layers does not change much after filtering or averaging the equations and the main bottleneck of the numerical simulation is not alleviated.

Accordingly, the standard rules in meshing boundary layer regions are that the first cell center should not have a wall-normal edge greater than $\approx \delta^+$ for RANS and $\approx 0.1\delta^+$ for LES and DNS. An additional requirement is that about 15–20 cells should be located inside the boundary layer thickness δ_{bl} whose value decreases as Re^b with $b = -0.5$ in laminar and $b \approx -0.2$ in turbulent flows (Schlichting, 1968).

Looking at Figure 8.1 it is clear that such resolutions can be attained only using body-fitted meshes by increasing the grid stretching, while Cartesian grids become impractical. Another disadvantage of the latter is that the crossing of curvilinear boundaries with perpendicular coordinate lines yields an extremely uneven distribution of wall-normal distances of the first external nodes, which further degrades the numerical solution (Figure 8.4a).

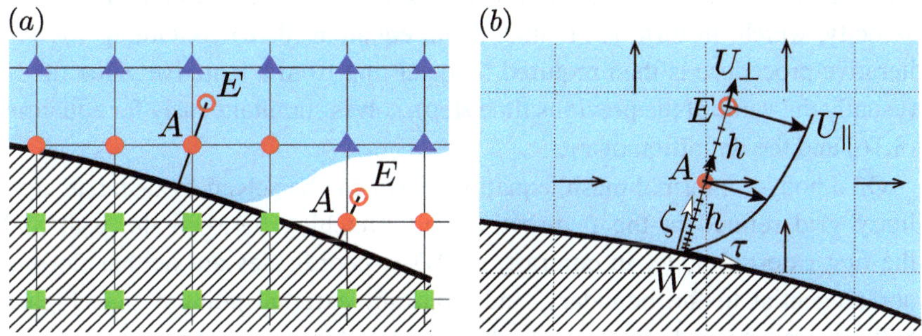

Figure 8.4 (a) Intersections of a curvilinear boundary with a Cartesian grid. Bullet points are color coded and visually distinguished as follows: internal points (green filled squares), first external points A (red filled circles), and fluid points (blue filled triangles). Points E, shown as red open circles, are used to apply the external boundary condition for the wall-model equation (8.16). (b) Detail of the setup for the solution of equation (8.16) for one of the first external points in a background grid with a staggered discretization.

It should be noted that, although wall resolution is a bottleneck for the application of IBMs to high-*Re* flow, the same problem is a serious limitation also for the simulations with body-fitted meshes; Tennekes and Lumley (1972) had already put forward the idea of using asymptotic solutions (wall functions) to bridge the near-wall region of turbulent flows to avoid the fine discretization needed therein.

The original idea was based on the "law of the wall," which provides a static, semi-empirical function for $\mathbf{U}(\zeta)$ in the turbulent boundary layer. This function is used to impose the velocity at the first external node ζ_1, which can therefore be positioned farther from the boundary.

Balaras et al. (1996) and Cabot and Moin (2000) proposed a dynamic version of the "law of the wall" consisting of a simplified boundary layer equation for the near-wall tangential velocity $U_\parallel$ which can provide the value at ζ_1 or, as an alternative, the wall shear stress. The equation reads

$$\frac{\partial}{\partial \zeta}\left[(\nu + \nu_T)\frac{\partial U_\parallel}{\partial \zeta}\right] = F \qquad \text{with} \qquad F = \frac{\partial U_\parallel}{\partial t} + \frac{\partial U_\parallel U_\parallel}{\partial \tau} + \frac{\partial U_\parallel U_\perp}{\partial \zeta} + \frac{\partial P}{\partial \tau},$$

$$(8.16)$$

with ζ and τ the local normal and tangential coordinates, respectively (for the ease of representation, only the two-dimensional equation is written).

The eddy viscosity ν_T is obtained by a mixing length model with near-wall damping (van Driest, 1956): $\nu_T = \nu\kappa\zeta^+(1-e^{-\zeta^+/\mathscr{A}})^2$, with $\kappa = 0.4$ and $\mathscr{A} = 19$. Here, $\zeta^+ = \zeta/\delta^+$ is the distance from the wall in viscous units computed from the instantaneous local friction velocity u_τ; see also the end of Section 10.6. It must be noted that the calculation of ν_T needs ζ^+, which relies on the friction velocity, which, in turn, is obtained from equation (8.16) containing ν_T. An iterative procedure is then required that, starting from a tentative value of u_τ (usually the value at the previous time step), solves simultaneously for equation (8.16) and the definition of ν_T.

On a body-conformal mesh, equation (8.16) can be solved by using an auxiliary grid refined in the ζ direction and stretching between the wall and the first external node. In contrast, in IBMs, coordinates are neither normal nor tangential to the boundary, and from any external point A (see Figure 8.4b) a wall-normal ray is cast; this is extended on the opposite side by the same distance h. Since point E generally does not coincide with a grid node, interpolations are needed to compute its tangential and normal velocity components, which are used as boundary conditions for equation (8.16) integrated over a one-dimensional refined mesh between E and W. When the solution is obtained, $U_\parallel$ at point A is available and its projection in the appropriate direction constitutes the velocity value for the IB forcing and reconstruction.

It is worth noting that the right-hand side F of (8.16) contains both ζ and τ derivatives and involves also $U_\perp$ and P, in addition to $U_\parallel$; thus, its evaluation entails considerable computational overhead. In order to avoid this complication, the model (8.16) is often employed imposing $F = 0$ and it is referred to as the equilibrium stress model, which has proven to perform quite well both with body-fitted meshes and IBMs.

It has long been thought that the model of equation (8.16) would perform even better if the full term F were computed and that the main reason for using the simple equilibrium stress model is that it reduces to a local ordinary differential equation rather than the full partial differential equation. There have also been attempts to include single terms of F, especially $\partial P/\partial\tau$, which gives the streamwise pressure gradient and the related boundary layer dynamics. However, Larsson et al. (2016) have shown that the wall model captures only the inner layer whose dynamics are much faster than the outer counterpart and it is thus mostly in equilibrium. Furthermore, the terms of F are evaluated in the outer layer where, even if they are individually large, they are in balance, thus yielding a vanishing F and justifying the good performance of the equilibrium stress model.

An example of a simulation using IBMs with LES and wall models is that of Tessicini et al. (2002), who considered the flow past a hydrofoil at high Reynolds number, already investigated experimentally by Blake (1975) and numerically by Wang and Moin (2000) using a body-fitted mesh and LES turbulence model.

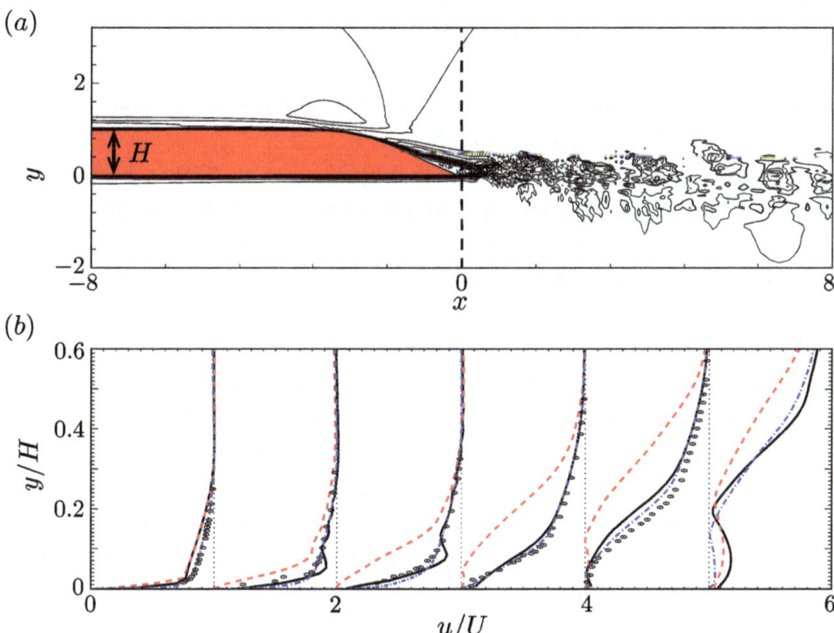

Figure 8.5 (a) Instantaneous snapshot of a vertical section of the flow past a hydrofoil trailing edge. Contours ($[-0.2 : 1.2]$ with increment 0.08) show the instantaneous streamwise velocity. The vertical dashed line indicates the section $x = 0$ at the trailing edge. (b) Mean velocity magnitude at $x/H = -3.125$, -2.125, -1.625, -0.625 and 0, from left to right. Each profile is shifted rightward by one unit for clarity of representation. Symbols: experiment by Blake (1975); blue — · — body-fitted wall-resolved full LES from Wang and Moin (2000); red − − − IBM LES without wall model; black ——— IBM LES with wall model (8.16) with $F = 0$. Figures adapted from Tessicini et al. (2002).

The results, reported in Figure 8.5, show that the smooth curvature of the suction side of the trailing edge determines a flow separation whose nature is completely viscous and requires the correct description of the boundary layers dynamics to be captured. The simulation with IBM on a Cartesian mesh could achieve a wall resolution ranging between $2\delta^+$ and $60\delta^+$, which is the reason for the misprediction of the flow separation. On the other hand, the same method when complemented with the wall model (8.16) in the equilibrium stress version $F = 0$, yielded much better results that compared well with the experiment and the simulation with the body-conformal mesh.

Since these initial attempts, several improvements and generalizations have been proposed, and they are detailed and validated in several papers.

Among many, we mention the method suggested by Kang (2015) who solves the equilibrium stress model near the immersed boundary to obtain a turbulent viscosity used to compute the total shear stress at the wall that constitutes the boundary condition for the external flow.

A more sophisticated approach is that by Yang et al. (2015), based on the integral boundary layer equation of von Karman and Pohlhausen; in this case, rather than numerically integrating a simplified boundary layer equation, a velocity profile must be assumed and its coefficients determined by matching physical constraints. The authors showed that the model combines a Reynolds independent computational cost with the possibility of handling resolved and subgrid roughness, thus making it particularly appealing for applications like atmospheric boundary layers of flows over complex rough surfaces.

Finally, Bernardini et al. (2016) have developed a model which relied on an arrangement as in Figure 8.4b except for the distance AE that was fixed as the maximum among all external nodes in order to alleviate the strong dependence of the numerical results of equation (8.16) on the location of the first off-wall grid node, which is particularly annoying for IBMs.

8.5 Adaptive Mesh Refinement

As previously mentioned, the need for a refined mesh is localized mostly in the steep gradient regions of the flow variables, which usually occupy only a small volume fraction of the computational domain. Following the philosophy of IBMs, it seems therefore a natural step to *immerse* refined grid patches within a base discretization as in the sketch of Figure 8.2b to resolve the equations on fine meshes only where necessary. This strategy has already been pursued in several contexts not involving IBMs; for example, Benek et al. (1986) explain the difficulties in obtaining usable body-fitted grids for three-dimensional complex objects and the possibility to alleviate them by decomposing the domain into subdomains and meshing each one with grids partially overlapping with their neighbors. This approach, referred to as "chimera grids" or "embedded grids" or "overset grids," relies on the idea of sharing information among bordering subdomains to obtain smooth fields over the entire domain.

A different approach is the one described in Aftosmis et al. (1998) in which a unique Cartesian mesh is recursively split, by the OCTREE approach, in selected subvolumes to refine the mesh where needed. The body can be described either through the stairstep surface created by the tiny Cartesian elements or the interface cuboids can be locally cut by the surface to obtain a smooth geometry. In fact, these can be considered IBMs although the data structure after the discretization is essentially unstructured, thus losing the ease and efficiency of structured meshes.

An interesting variation is that proposed by Durbin and Iaccarino (2002) who, starting from the finest grid level covering the entire volume, applied a recursive *coarsening* to eliminate grid nodes in selected regions. The main advantage of this approach is that all connectivity information of the final grid can be stored from the original structured mesh and also the initial ray-tracing step to tag the nodes can be performed using the structured data.

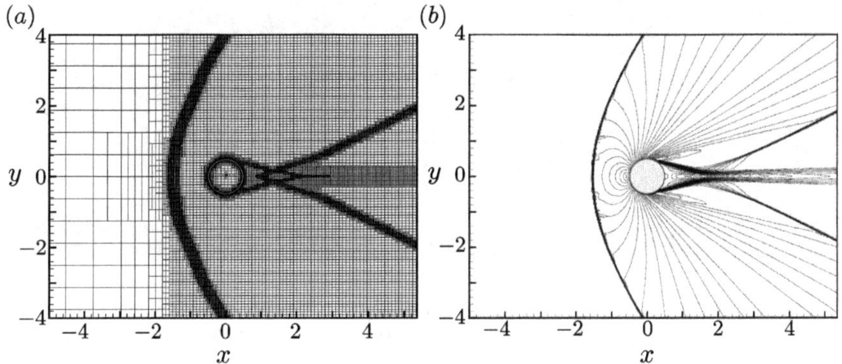

Figure 8.6 (a) Grid with adaptive mesh refinement for the two-dimensional supersonic flow around a circular cylinder. (b) Mach number contours for the steady flow at $M_\infty = 1.7$ and $Re = 2 \times 10^5$. The red circle indicates the boundary of the immersed cylinder of unity diameter. Figures adapted from de Tullio et al. (2007).

Any of the aforementioned approaches can be applied not only next to wall regions but also in the bulk if sharp flow gradients are present therein; this is certainly the case for shear layers in a body wake or shock waves forming ahead of it in a supersonic flow. An example is shown in Figure 8.6 for the two-dimensional flow around a circular cylinder at free-stream Mach number $M_\infty = 1.7$ and $Re = 2 \times 10^5$; thanks to an aggressive adaptive mesh refinement, not only the thin shear layer at the wall but also the complex system of shocks and waves can be captured. In this example, owing to the compressible nature of the flow, the authors have employed a TVD third-order-accurate upwind scheme capable of controlling the numerical oscillations of the variables across the discontinuities.

Also, the high-Re RANS simulations with adaptive grid refinement considered in Kalitzin et al. (2003) relied on upwind discretizations to stabilize the solution, while Capizzano et al. (2019) used a similar approach in combination with a matrix artificial diffusion.

Unfortunately, direct numerical simulations cannot be performed introducing numerical dissipation (Mahesh and Moin, 1998) as the fundamental energy transfer mechanisms and delicate balances are spoiled. Similar requirements have been found for LES in which the only additional dissipation has to come from the subgrid-scale model and central nondissipative discretizations are mandatory (Mittal and Moin, 1997).

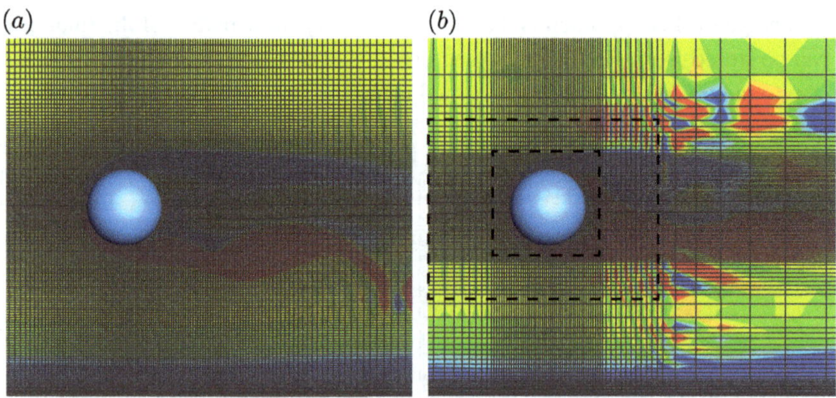

Figure 8.7 (a) LES with dynamic Smagorinsky model for the flow around a sphere at a distance $D = 4R$ from a solid wall at $Re = UR/\nu = 1000$. Contours in the vertical symmetry section of out-of-plane vorticity (values range from -1 to 1 from blue to red. Overlaid light gray lines are those of the actual Cartesian mesh). (b) The same as (a), except for the mesh which has been abruptly coarsened at the boundaries of the two windows indicated by the thick dashed lines. Note that the wall resolution at the sphere surface and at the lower solid boundaries is the same as for panel (a).

The effects of sudden jumps in the grid spacing, as they occur using adaptive mesh refinement, are shown in Figure 8.7, reporting the comparison of two otherwise identical LES simulations except for the mesh: In the first the spacing varies smoothly in space while in the second some coordinate lines have been deleted to create discontinuities in the metrics. It can be observed that the latter

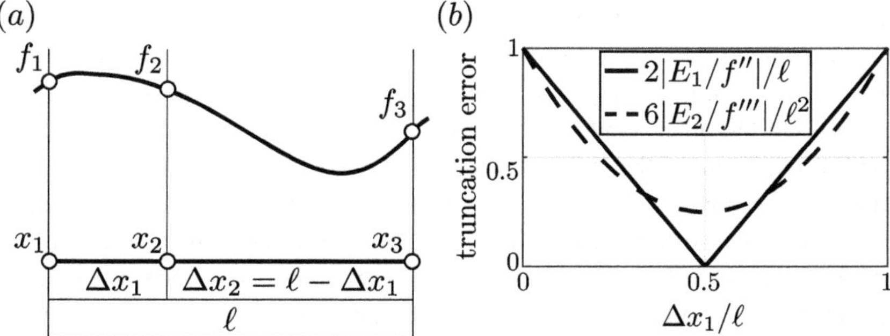

Figure 8.8 (a) One-dimensional function with nodal values $f(x_1) = f_1$, $f(x_2) = f_2$ and $f(x_3) = f_3$. The grid spacing is defined as $\Delta x_1 = x_2 - x_1$ and $\Delta x_2 = x_3 - x_2$ with the constraint $\Delta x_1 + \Delta x_2 = \ell$. (b) First- and second-order truncation error (normalized) of the central finite difference scheme as a function of $\Delta x_1/\ell$.

are the source of spurious wavy flow structures, which tend to spread in the computational domain and spoil the whole numerical solution.

The reason for this behavior is understood on account of the following simple argument: Consider the generic function $f(x)$ in Figure 8.8a defined on an interval of fixed length ℓ with nodal values f_1, f_2 and f_3 at the grid points x_1, x_2 and x_3, respectively, whose distance is $\Delta x_1 = x_2 - x_1$ and $\Delta x_2 = x_3 - x_2$ (with $\Delta x_1 + \Delta x_2 = \ell$). The forward and backward Taylor-series expansions of $f(x)$ about the central node x_2 yield:

$$f_1 = f_2 - f'(x_2)\Delta x_1 + \frac{1}{2}f''(x_2)\Delta x_1^2 - \frac{1}{6}f'''(x_2)\Delta x_1^3 + O\left(\Delta x_1^4\right),$$
$$f_3 = f_2 + f'(x_2)\Delta x_2 + \frac{1}{2}f''(x_2)\Delta x_2^2 + \frac{1}{6}f'''(x_2)\Delta x_2^3 + O\left(\Delta x_2^4\right), \tag{8.17}$$

which, by subtracting the first from the second, gives

$$f'(x_2) = \frac{f_3 - f_1}{\ell} \underbrace{- \frac{1}{2}f''(x_2)(\ell - 2\Delta x_1)}_{E_1} +$$
$$\underbrace{- \frac{1}{6}f'''(x_2)\left(\ell^2 - 3\ell\Delta x_1 + 3\Delta x_1^2\right)}_{E_2} + O\left(\Delta x_1^3, \Delta x_2^3\right). \tag{8.18}$$

The first term on the right-hand side is the central finite-difference first derivative followed by a first- and a second-order truncation errors E_1 and E_2.

Figure 8.8b reports both errors as a function of the grid size Δx_1, showing that both are symmetric about $\Delta x_1 = \Delta x_2 = \ell/2$. $|E_1|$, in particular, depends linearly on $\ell - 2\Delta x_1 = \Delta x_2 - \Delta x_1$, which is the size difference between two neighboring mesh elements. For a uniform discretization ($\Delta x_1 = \Delta x_2 = \ell/2$),

E_1 is null and the central finite-difference first derivative is formally second-order accurate. In contrast, if the grid is stretched the same expression is only first-order accurate as $\Delta x_1 \neq \Delta x_2$ yields $E_1 \neq 0$ and the second-order accuracy is lost. On the other hand, the second-order truncation error $|E_2|$ is quadratic with a minimum at the symmetry axis $\Delta x_1 = \Delta x_2 = \ell/2$ where $E_1 = 0$, thus there is a region around this point where $|E_2| > |E_1|$ and even if the method is formally first-order accurate it behaves second-order. This justifies the rule of thumb that grid spacing must be varied smoothly, and node distributions prescribed by continuous analytical functions have to be preferred to other rules.

Coming back to adaptive grid refinement, it can be considered as an extreme case of non-uniform mesh with $|\Delta x_2 - \Delta x_1| \approx \ell$; in this case, not only is the derivative first-order but the discretization error becomes dominant, acting as a forcing on the equations and, unless combined with dissipative schemes, it affects the numerical solution over the entire domain (Verzicco, 2023).

9

Advanced IBM Applications

Looking back at the various IBM implementations discussed so far, it is clear that their essence reduces to the imposition, on nonboundary mesh nodes, of velocity conditions. These can either be assigned directly as boundary values or obtained from separate physical models describing additional phenomena, like the motion of rigid objects or the local equilibrium of deformable solids.

It is then clear that the same approach can be applied to any partial differential equation to alter the governed variable over the immersed interfaces and so avoid the use of body-conformal meshes. Indeed, Cao et al. (2023) and Lee and Liu (2023) have used immersed boundary methods to solve Maxwell's equations, Schillinger et al. (2016) for Kirchhoff plates and three-dimensional elasticity problems, Codony et al. (2019) to study piezoelectric phenomena in solids and Carraturo et al. (2021) for 3D-printing and additive manufacturing applications.

In the following, we will describe some possible extensions of immersed boundary methods that also involve fluid motion.

9.1 Heat Transfer

We consider a rigid body B, whose position is fixed in space, subjected to a uniform flow as in Figure 9.1; the problem of imposing the appropriate velocity boundary conditions along ∂B, and possibly assigning the velocity inside the body, have already been extensively discussed in the previous chapters, so here we focus on the case in which fluid and body temperatures do not match.

Several scenarios are possible, depending on the thermal properties of the body B: If the heat capacity of the solid object is much larger than that of the fluid, or if the body temperature is actively maintained uniform and constant, then it can be assumed also that the body surface is isothermal, say at T_b, and the problem would be that of assigning to all the first external points (red bullets of

155

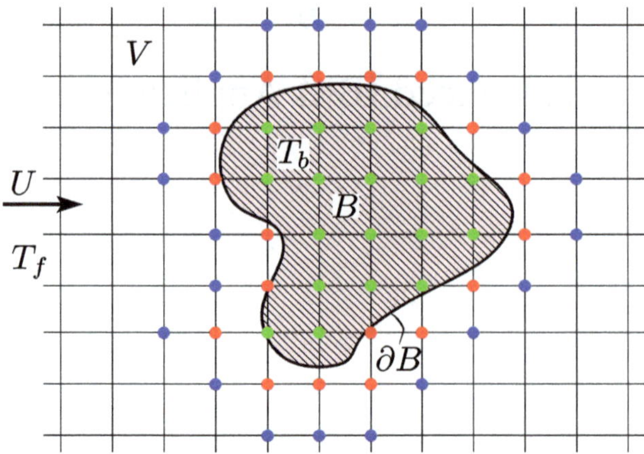

Figure 9.1 Sketch of a rigid, fixed body B having temperature T_b and subjected to a uniform flow with velocity U and temperature T_f in a domain V.

Figure 9.1) a temperature such that T_b is imposed on ∂B. As an option, it is also possible to force the temperature T_b for all internal points (green bullets) and both results can be obtained using one of the forcings described in Chapter 4. As an example, if the direct forcings of Section 4.2 are employed, the continuous temperature equation becomes

$$\frac{\partial T}{\partial t} + \mathbf{u} \cdot \nabla T = \kappa \nabla^2 T + h, \tag{9.1}$$

with κ the thermal diffusivity of the fluid and h the volumetric forcing term. It can be recast in time discrete form as

$$\frac{T^{n+1} - T^n}{\Delta t} = rhs^{n+1/2} + h^{n+1/2}, \tag{9.2}$$

where $rhs^{n+1/2}$ contains convective and diffusive terms, to be solved for the forcing $h^{n+1/2}$ once the temperature has been altered:

$$h^{n+1/2} = \frac{T_b^{n+1} - T^n}{\Delta t} - rhs^{n+1/2}. \tag{9.3}$$

Here $T_b^{n+1} \equiv T_b$ for all internal points, while for the first external points T_b^{n+1} is obtained by one of the boundary reconstruction procedures already detailed in Section 4.2

Another noticeable case is that of a body for which the specific heat flux is prescribed at its surface $\lambda \nabla T \cdot \mathbf{n} = \dot{q}_W$, with λ the thermal conductivity of the fluid, which is equivalent to imposing the value of the wall-normal derivative of temperature $\partial T / \partial n = \dot{q}_W / \lambda$. The possibility of imposing derivatives of the

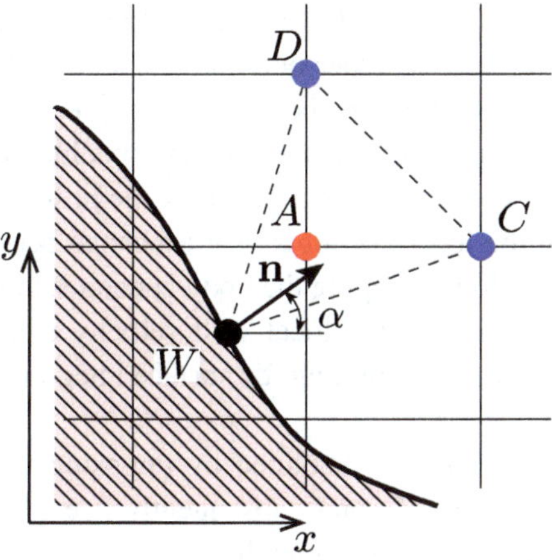

Figure 9.2 Sketch for temperature interpolation at a boundary point.

forced variable at the boundary, already explained in Section 4.2.1, makes the previous task a simple one.

In fact, assuming that the temperature field next to the solid boundary is smooth enough to be described by a linear function, we can write $T(x, y) = a + bx + cy$ with a, b and c unknown coefficients. From Figure 9.2, we see that the wall-normal vector in W can be assigned through its components $\mathbf{n} = (\cos\alpha, \sin\alpha)$, and the temperature derivative in that direction can be expressed as $\nabla T \cdot \mathbf{n} = b\cos\alpha + c\sin\alpha$. On the other hand, since D and C are not first external points, their temperature is known from the equations; thus the following system can be written:

$$\begin{cases} a + bx_D + cy_D = T_D, \\ a + bx_C + cy_C = T_C, \\ b\cos\alpha + c\sin\alpha = \dot{q}_W/\lambda, \end{cases} \tag{9.4}$$

which can be solved for the unknown coefficients a, b and c. These, in turn, are used to evaluate the temperature at the first external node $T_A = a + bx_A + cy_A$, which is plugged into an expression like (9.3) to enforce the desired temperature boundary condition.

In a three-dimensional arrangement, the temperature would be described by a function $T(x, y, z) = a + bx + cy + dz$ with the triangle WDC of Figure 9.2 which becomes a tetrahedron with three external points and the system (9.4) consisting of four equations in the unknowns a, b, c and d.

Finally, if the triangle of Figure 9.2 is extended in the fluid to include more external points, it is possible to parametrize the local temperature $T(x, y)$ by higher-order polynomials or to enforce additional constraints; nevertheless, this increases the stencil over which T_A is evaluated and decreases its locality. In practice, only first- or second-order polynomials are employed, because more complex functions do not improve the accuracy or precision of the solution.

It is worth mentioning that, regardless of the specific boundary conditions, the aforementioned procedure is often used for the postprocessing of the temperature field when the local or global heat exchange between solid and fluid is needed. In fact, the flow sweeping the body generates kinematic and thermal boundary layers which drive the local wall velocity and temperature gradients, respectively, in turn determining, by surface integration, friction and heat transfer coefficients.

A more complex problem is obtained by imposing that the body is only *initially* at the temperature T_b and subsequently it evolves according to the equilibrium with the surrounding flow. In this case, temperatures in the body and in the fluid influence each other and this is referred to as a conjugate heat transfer problem. In principle, the domain V of Figure 9.1 could be split into body B and fluid $V_f = V - B$ domains to solve the heat conduction equation in the former and the Navier–Stokes with temperature equation in the latter. The two problems would be coupled through the continuity across ∂B of temperature, $T|_{\text{solid}} = T|_{\text{fluid}}$, and of the heat flux which reads $\lambda_b \nabla T \cdot \mathbf{n}|_{\text{solid}} = \lambda_f \nabla T \cdot \mathbf{n}|_{\text{fluid}}$.

Taking advantage of the flexibility of immersed boundary methods, both problems can be solved within a single set of equations without explicitly imposing the earlier boundary conditions. In fact, we can write

$$\frac{\partial \mathbf{u}}{\partial t} + \mathbf{u} \cdot \nabla \mathbf{u} = -\frac{\nabla p}{\rho} + \nu \nabla^2 \mathbf{u} + \mathbf{f}, \qquad \nabla \cdot \mathbf{u} = 0, \tag{9.5}$$

$$\frac{\partial T}{\partial t} + \mathbf{u} \cdot \nabla T = \frac{1}{\rho C} \nabla \cdot (\lambda \nabla T) + h,$$

valid over the whole domain V. The volumetric forcing $\mathbf{f}$ of the momentum equation (9.5) can be such to set to zero[1] the velocity inside the solid body and to impose no-slip boundary conditions on ∂B (if the body B is not at rest, a solid-body motion can be imposed). It should be remarked that the second of equations (9.5) automatically satisfies the continuity of temperature and heat flux at the interfaces of materials with different thermal properties (where the conjugated heat transfer occurs). Therefore, in the temperature equation h will be nonzero only within regions where the temperature, or its gradient, is imposed; in the remaining domain (including the fluid/solid interface) the equation is solved using for ρ, C and λ the local values of the medium of density, specific heat and thermal conductivity, respectively.

---

[1] It is worthwhile noting that for this problem, differently from Section 4.3, the forcing to zero of the velocity of the inner point of the body is not optional. In fact, only heat conduction occurs inside the body and this is obtained from the second of equation (9.5) by zeroing the convective terms.

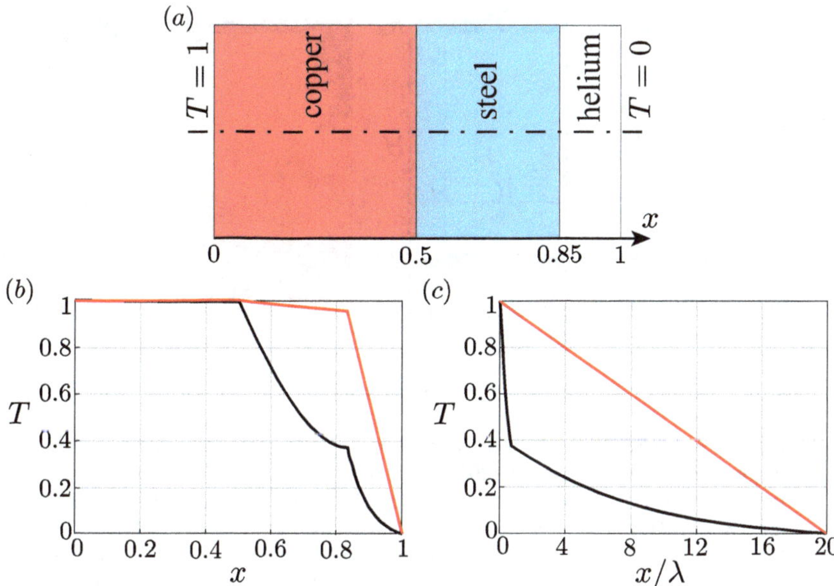

Figure 9.3 (a) Setup of a thermal conduction problem with three contiguous materials joined along the x-direction. The following material properties have been assumed, respectively for gaseous helium, stainless steel and copper (at the temperature of $T = 4.5$ K): $\lambda_{\mathrm{He}} = 0.0087$ W/(m K), $(\rho C)_{\mathrm{He}} = 3190$ J/(m^3 K), $\lambda_{\mathrm{Fe}} = 0.4$ W/(m K), $(\rho C)_{\mathrm{Fe}} = 30\,000$ J/(m^3 K) and $\lambda_{\mathrm{Cu}} = 1000$ W/(m K), $(\rho C)_{\mathrm{Cu}} = 890$ J/(m^3 K). (b) Temperature profile $T(x, t)$ during the initial transient (black line at $t = 5$ time units) and at steady state (red line at $t = 40$). (c) The same as (b) but for plotted against the scaled abscissa $x/\lambda(x)$. (Adapted from Verzicco (2002).) © Cambridge University Press, reproduced with permission.

This procedure has been tested for a sample of three contiguous materials, copper, stainless steel and gaseous helium, of different lengths subjected to a temperature difference at the extrema (Figure 9.3a). Any convective motion has been suppressed by setting $\mathbf{u} = \mathbf{0}$ everywhere, thus the problem is that of pure thermal conduction in a medium with inhomogeneous thermal properties. The steady-state solution of the temperature equation is a piecewise linear function within each material, as shown in Figure 9.3b. Furthermore, using the condition at the interfaces $\lambda_i \nabla T \cdot \mathbf{n}|_i = \lambda_j \nabla T \cdot \mathbf{n}|_j$, with i and j the subscripts indicating two adjacent materials, we can see that when plotted against the abscissa $x/\lambda(x)$ the steady-state solution must become a unique straight line: The absence of any slope discontinuity in the stationary solution of Figure 9.3c confirms that the proposed procedure yields correct results.

The set of equations (9.5) can be employed for more complex configurations, like that of Figure 9.4a in which multiple materials with different thermal properties are assembled to build a closed cell containing a fluid (Novec 7000) in

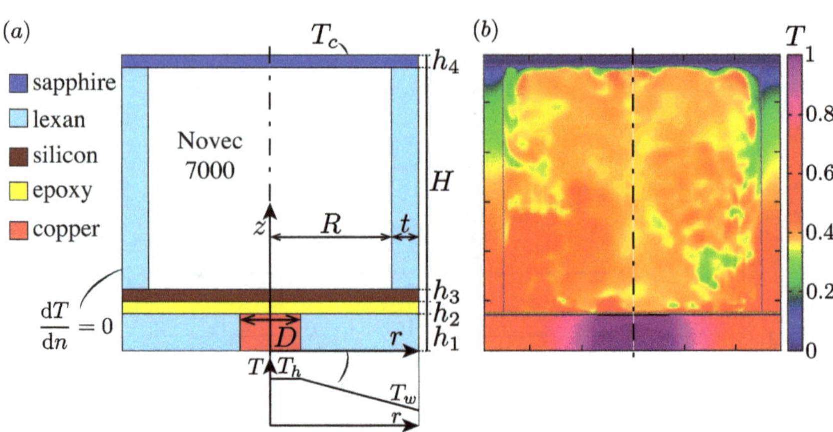

Figure 9.4 (a) Setup of a thermal convection cell made from multiple materials inspired from Hoefnagels et al. (2017). The upper boundary is isothermal (T_c) on the "dry" side and no-slip on the "wet" side. The lower plate is isothermal (T_h) for $0 \leq r \leq D/2$ and isothermal with a temperature varying between T_h and T_w for $D/2 < r \leq R + t$ ($T_w = 0.53T_h$ in the present example). The plate is no-slip on the wet side. The sidewall is adiabatic on the "dry" side and no-slip on the wet side. The scaling quantities are: The main length is $H = 8.84$ cm, and the temperature difference $\Delta = T_h - T_c = 1.8$ K; with the fluid (Novec 7000) properties $\rho_f = 1400$ kg/m^3, $C = 1300$ J/(kg K), $\lambda_f = 0.075$ W/(m K), $\alpha = 0.00219$ K^{-1} and $\nu = 3.2 \times 10^{-7}$ m^2/s, the resulting Rayleigh number is $Ra \simeq 2 \times 10^9$. The other geometrical dimensions are: $h_1/H = 0.1414$, $h_2/H = 0.0051$, $h_3/H = 0.0060$, $h_4/H = 0.037$, $D/H = 0.286$, $R/H = 0.5028$ and $t/H = 0.072$. The thermal properties of the solid materials are: sapphire $\lambda_{sap}/\lambda_f = 306.6$, $(\rho C)_{sap}/(\rho_f C) = 1.66$; epoxy $\lambda_{epo}/\lambda_f = 2.37$, $(\rho C)_{epo}/(\rho_f C) = 0.714$; Lexan $\lambda_{lex}/\lambda_f = 2.8$, $(\rho C)_{lex}/(\rho_f C) = 0.824$; copper $\lambda_{cop}/\lambda_f = 5333$, $(\rho C)_{cop}/(\rho_f C) = 1.9$. (b) Instantaneous snapshot of the temperature field for a flow at $Ra \simeq 2 \times 10^9$ and $Pr = 7.76$. Three-dimensional direct numerical simulation performed in cylindrical coordinates on a mesh of $257 \times 193 \times 381$ nodes in the azimuthal, radial and vertical directions, respectively.

which a turbulent flow is induced by temperature differences. An instantaneous temperature field is reported in Figure 9.4b showing the distribution in the fluid and in the solid domains.

9.2 Mass Transport Across Moving Interfaces

In many contexts, a small parcel of a fluid evolves in a medium; this also contains the first fluid, possibly in a different phase. This is the case of a water droplet moving in air with moisture, an air bubble in non-degassed water or hydrogen bubbles next to an electrode during water electrolysis. As a model problem, we can consider a spherical gas bubble of radius R in a liquid in which the gas is dissolved, see Figure 9.5. Let C be the local concentration of the gas in the liquid phase; it must satisfy the convection–diffusion equation

$$\frac{\partial C}{\partial t} + \mathbf{u} \cdot \nabla C = d\nabla^2 C + h, \tag{9.6}$$

with d the gas diffusivity in the liquid and h the volumetric forcing term to be used at the gas–liquid interface to satisfy the boundary conditions. We indicate by P_0 the gas pressure in the bubble, which depends on the local value of the pressure in the liquid and on the Young-Laplace pressure produced by the surface tension, and rely on Henry's law to determine the concentration boundary condition at the interface $C_{sat} = k_c P_0$, where k_c is the specific Henry constant for the gas in the liquid.

The solution of equation (9.6) in the liquid with the aforementioned boundary conditions yields the concentration field from which its gradient at the bubble surface is computed and then the net mass flux through the boundary via

$$\dot{m}_C = \int_{\partial B} d\nabla C \cdot \mathbf{n} dA. \tag{9.7}$$

This mass flux, in turn, alters the bubble volume and, assuming the bubble retains its spherical shape, it allows us to determine the rate of change of the bubble radius

$$\frac{dR}{dt} = \frac{\mathcal{R}T_0}{P_0} \frac{1}{4\pi R^2} \dot{m}_C, \tag{9.8}$$

with T_0 the temperature of the gas in the bubble and $\mathcal{R}$ the constant in the equation of state for perfect gases.

The bubble will be subjected to pressure and viscous hydrodynamic loads, buoyancy force and its own weight, whose resultant we denote by $\mathbf{F}$; the position of its centroid can therefore be determined through the equation

$$\frac{d^2\mathbf{x}_B}{dt^2} = \frac{\mathcal{R}T_0}{P_0} \frac{3}{4\pi R^3} \mathbf{F}, \tag{9.9}$$

which yields the velocity of the bubble centroid $\mathbf{u}_B = d\mathbf{x}_B/dt$. The growth or shrinking of the bubble radius produces an additional velocity perpendicular to the interface $\mathbf{u}_\perp = (dR/dt)\mathbf{n}$; thus the boundary condition to be imposed in the momentum equation at the immersed boundary is $\mathbf{u}_{\partial B} = \mathbf{u}_B + \mathbf{u}_\perp$, which can be achieved by one of the forcings described in Chapter 4.

It is worthwhile mentioning that the mass flux across the immersed interface of equation (9.7) is purely diffusive and does not involve any velocity of the liquid phase perpendicular to the surface. On the other hand, as the bubble grows or shrinks, the velocity $\mathbf{u}_\perp = (dR/dt)\mathbf{n}$ violates the continuity equation $\nabla \cdot \mathbf{u} = 0$, which must be solved over the whole domain (also inside the immersed object, see Figure 9.5).

Following Lupo et al. (2020) and Sapahi et al. (2024), the problem can be avoided by solving a modified continuity equation

$$\nabla \cdot \mathbf{u} = \ell, \tag{9.10}$$

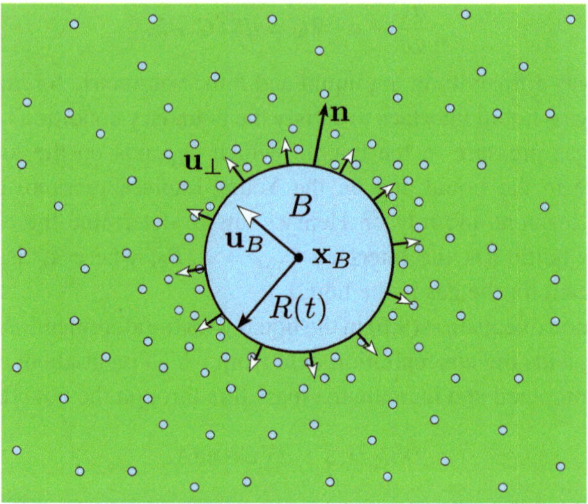

Figure 9.5 Sketch for the problem of mass transport across a moving interface. In the drawing it is assumed that $dR/dt > 0$.

with ℓ a forcing term which is zero in the liquid phase and $\ell = (1/V)(dV dt) = (3/R)(dR/dt)$ inside the bubble: The latter expression comes from the assumption that the bubble shape remains spherical.

A recent application of the aforementioned technique, including electrochemical reactions at the electrode and buoyancy-driven convective motion in the liquid phase, is presented in Sapahi et al. (2024) and a representative result is shown in Figure 9.6; in that paper, a dilute sulfuric acid solution is considered as the electrolyte and the water-splitting process, induced by electrical current at the cathode, is analyzed in detail to study hydrogen production.

9.3 Diffusiophoresis at Reactive Surfaces

Particles in a fluid can be produced by chemical reaction, electrolysis, dissolution or biological processes, among many possible mechanisms. When the particle size is at the colloidal scale, the movement occurs in response to their concentration gradient and the phenomenon is referred to as diffusiophoresis.

A simple setup to illustrate the basic dynamics is that of Figure 9.7a with a flat surface aligned with the x-direction whose catalytic properties set the wall-normal concentration according to

$$d\frac{\partial C}{\partial y}\bigg|_W = \alpha, \tag{9.11}$$

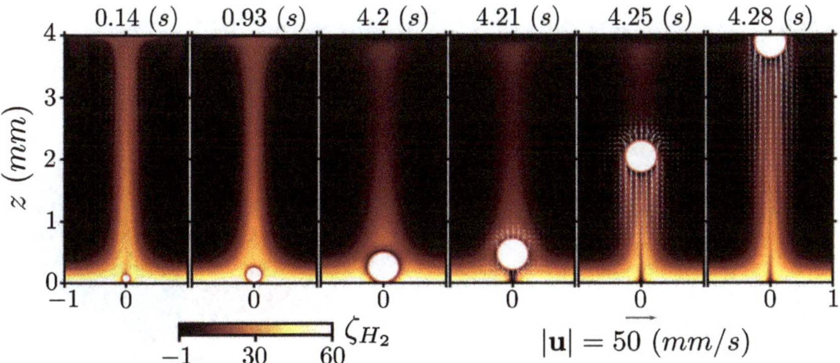

Figure 9.6 Instantaneous snapshots of a hydrogen bubble forming at an electrode ($z = 0$) with current density $I = 10^2$ A/m^2. The bubble is initiated at $t = 0$ s with a radius $R = 25$ μm and breaks off the electrode when $R \geq 0.25$ mm. The isosurface is color-mapped by hydrogen supersaturation level $\zeta_{H_2} = C/C_{sat} - 1$, with a legend indicating the corresponding values. The vectors indicate the velocity field induced by the growth and rise of the bubble in the electrolyte. (Adapted from Sapahi et al. (2024).) © Cambridge University Press, reproduced with permission.

where d is the diffusivity of the reaction product in the fluid and α is a measure of the reaction activity at the surface ($\alpha > 0$ if C is produced, $\alpha < 0$ if C is consumed).

Possible inhomogeneities at the catalytic plane or flow instability can also generate a wall-parallel component of the gradient; this produces a slip velocity at the wall given by

$$\mathbf{u} \cdot \hat{\mathbf{x}} = M \left. \frac{\partial C}{\partial x} \right|_W, \tag{9.12}$$

with M the phoretic mobility which can be positive or negative depending on the specific solute–surface interaction. It is worth noting that this relation should be considered as an effective boundary condition since there is a thin layer next to the wall (of the order of Debye length for ions or charged particles) where complex microscopic phenomena occur and short-range forces dominate the equilibrium (Anderson, 1989): The slip velocity is generated only outside this layer.

The wall-normal velocity component is instead zero, as should be expected from the no-penetration condition ($\mathbf{u} \cdot \hat{\mathbf{y}} = 0$).

Figure 9.7b shows a more popular general setup, a Janus spherical particle, in which only half of the surface is catalytic while the other half is inert. The uneven surface distribution of reaction activity causes concentration boundary conditions at the immersed interface

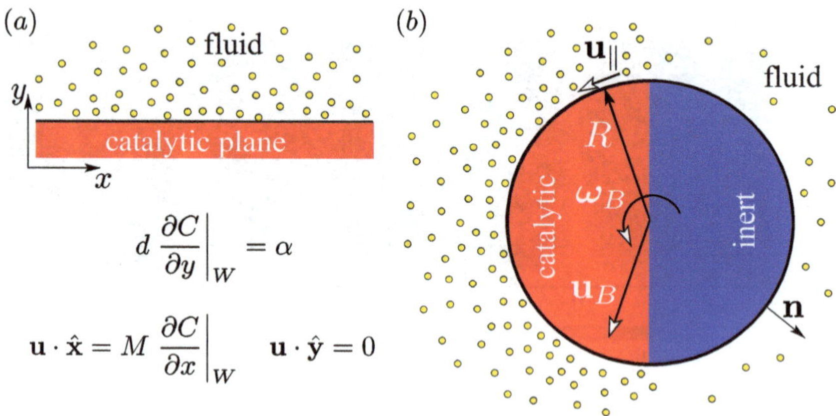

Figure 9.7 (a) Basic mechanisms of diffusiophoresis at a catalytic flat surface. (b) Diffusiophoresis at a Janus particle.

$$\nabla C \cdot \mathbf{n}|_W = 0, \text{ at the inert surface;} \qquad \nabla C \cdot \mathbf{n}|_W = \frac{\alpha}{d}, \text{ at the catalytic surface,} \tag{9.13}$$

which are enforced in an equation like (9.6) through the immersed boundary forcing h to determine the concentration field in the fluid phase. From the latter, the slip velocity at the surface is computed via

$$\mathbf{u}_\parallel = M(\mathbf{I} - \mathbf{nn}) \cdot \nabla C|_W, \tag{9.14}$$

where $\mathbf{I}$ is the identity tensor.

Indicating by $\boldsymbol{\tau}$ the viscous stress tensor, the hydrodynamic force and torque on the particle can be computed by using

$$\mathbf{F}_B = \int_{\partial B} (-p\mathbf{I} + \boldsymbol{\tau}) \cdot \mathbf{n} dS, \qquad \mathbf{T}_B = \int_{\partial B} \mathbf{R} \times [(-p\mathbf{I} + \boldsymbol{\tau}) \cdot \mathbf{n}] dS, \tag{9.15}$$

the torque being evaluated with respect to the sphere center. These loads, possibly complemented with other components like the weight of the particle and buoyancy $\mathbf{F}_e$, are the input for the dynamic equations

$$m_B \frac{d\mathbf{u}_B}{dt} = \mathbf{F}, \qquad I_B \frac{d\omega_B}{dt} = \mathbf{T}_B, \tag{9.16}$$

where m_B and I_B are the mass and moment of inertia of the particle, respectively, ω_B is its angular velocity and $\mathbf{F} = \mathbf{F}_B + \mathbf{F}_e$ is the total force. Note that here the second of (9.16) contains only $\mathbf{T}_B$ because for the spherical Janus particle buoyancy and gravity do not produce extra moment. For more general bodies, that contribution should also be considered.

After these steps, the absolute velocity vector at any point of the immersed surface can be computed through

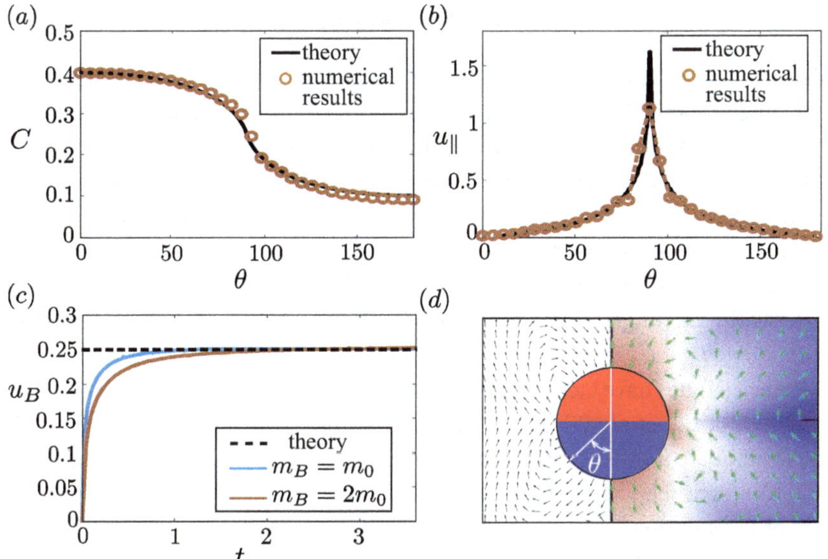

Figure 9.8 Comparison of numerical results with theoretical predictions by Golestanian et al. (2007) for a spherical Janus particle at $Sc = v/d = 510$ and $Pe = M\alpha L/d^2 = 8.45\ 10^{-3}$; v is the kinematic viscosity of the fluid and L the particle diameter. (a) Scalar concentration over the particle surface; (b) slip-velocity; (c) particle translation velocity ($m_0 = \rho\pi L^3/6$ is the mass of a sphere with diameter L and the same density as the fluid, m_B is the mass of the particle); (d) velocity vectors in a symmetry plane obtained with IBM (left), compared with the results of de Graaf et al. (2015). (Adapted from Zhu et al. (2024).)

$$\mathbf{u}_{\partial B} = \mathbf{u}_B + \omega_B \times \mathbf{R} + \mathbf{u}_\| \tag{9.17}$$

and enforced in a momentum equation like (9.5) through the immersed boundary forcing **f**.

The procedure detailed here has been extensively described in Zhu et al. (2024) and validated by comparisons with similar results available in the literature. In Figure 9.8 we report some results for a single Janus particle of diameter $L = 1$ μm in 25 percent H_2O_2 in water which, in contact with the catalytic part of the surface, produces O_2. The setup is that proposed by de Graaf et al. (2015) where further details can be found.

In the limit of small Peclet number, $Pe = M\alpha L/d^2$ with L the particle diameter, Golestanian et al. (2007) have computed the surface concentration of the reaction product, local slip-velocity and translation speed of the particle, all of which show excellent agreement with the results obtained by an immersed boundary method.

9.4 Melting of Solids in a Liquid

Melting or solidification of a material in its liquid phase is a phenomenon with huge practical relevance in nature and technology: Melting of icebergs (Cenedese and Straneo, 2023), formation of rocks from molten lava (Griffiths, 2000), latent heat storage devices (Alva et al., 2018) or industrial production by metal casting (Luo et al., 2022) are only a few examples. The dynamics of the two phases are strongly coupled both through momentum and temperature as buoyancy affects the flow and determines the local heat transfer which, in turn, governs the melting/solidification phenomena through the latent and sensible heat balance.

Recent studies have tackled these problems using phase field methods (Hester et al., 2020; Liu et al., 2021; Zhang et al., 2022b), which have been efficiently run to simulate a variety of melting problems (Couston et al., 2021; Yang et al., 2023b) where the shape of the solid–liquid interface had to be determined as part of the solution.

The essence of the phase field method is to solve an additional equation for an Eulerian scalar field ϕ whose value is initially set to 1 in the solid and 0 in the liquid phases:

$$\frac{\partial \phi}{\partial t} = \hat{A}\left[\nabla^2\phi - \frac{1}{\beta^2}\phi(1-\phi)(1-2\phi+\hat{C}T)\right]. \tag{9.18}$$

Equation (9.18) depends on both physical and numerical parameters: β is the diffusive interface thickness, which is typically set to be the mean grid spacing, while $\hat{C}$ is the phase mobility parameter, accounting for the Gibbs–Thompson effect, altering the melting temperature depending on the local curvature of the interface. Finally, $\hat{A}$ is determined by $\hat{C}$ but also depends on flow and thermomechanical material properties.

Once equation (9.18) is solved together with the appropriate momentum and energy equations (see for example, Couston et al. (2021)), the solid–fluid interface is obtained from the solution at the isosurface $\phi = 0.5$, where $T = T_m$.

An example is shown in Figure 9.9, where a fluid (water) is heated from above at temperature T_{hot} and cooled from below at T_{cold} such that the melting temperature T_m is in between. Note that since ice melts at $T_m = 0\,°\text{C}$ and water has a density anomaly for $0\,°\text{C} \le T \le 4\,°\text{C}$, the configuration of Figure 9.9 can generate thermal convection above the melted ice.

It is worth mentioning that the phase field method already uses an immersed-boundary-related approach since, while the temperature field is solved simultaneously in the liquid and in the solid region, the velocity field in the latter ($\phi = 1$) is forced to zero following the same algorithm described in Section 9.1.

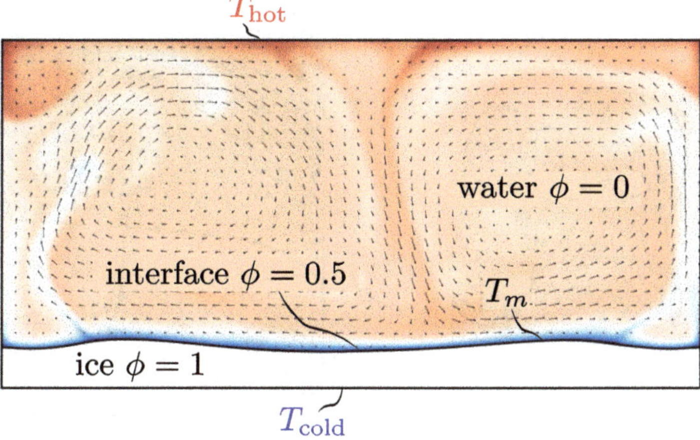

Figure 9.9 Instantaneous snapshot of temperature, velocity vectors and interface position for the melting of an ice block heated from above and cooled from below. Note that, owing to the density anomaly of water, a thermally driven convective motion can be generated in this configuration. The numerical simulation has been performed using a phase field method as described in Yang et al. (2023a). Adapted figure with permission from Yang et al. (2023a). Copyright 2023 by the American Physical Society.

An important feature of the problem of Figure 9.9 is that the solid phase is not moving through the liquid and the only interface velocity is that produced by the phase change (accordingly, equation (9.18) misses convective terms). On the other hand, a general melting problem would be as in the sketch of Figure 9.10, in which the solid body can move in its liquid phase while melting: Such an extension would be difficult for an equation like (9.18) since the computational nodes within the solid would change in time and cumbersome forcing procedures with interpolations and conservation issues would be needed.

Looking at Figure 9.10 we can write the heat flux balance at the liquid–solid interface to obtain the classical Stefan condition (Worster et al., 2000)

$$\mathbf{u}_\perp = \frac{\kappa}{\mathcal{L}C} \left[\nabla T|_S \cdot \mathbf{n}^+ - \nabla T|_L \cdot \mathbf{n}^- \right], \tag{9.19}$$

where κ, $\mathcal{L}$ and C are the thermal diffusivity, latent heat and specific heat, respectively, while $\nabla T|_S$ and $\nabla T|_L$ are temperature gradients at the interface evaluated in the solid and liquid phases.

Following a scheme like that described by equations (9.15)–(9.16), the motion of the solid object is determined through the velocity of its centroid $\mathbf{u}_B$ and the angular velocity ω_B; thus the total velocity of any boundary point can be computed as

$$\mathbf{u}_{\partial B} = \mathbf{u}_B + \omega_B \times \mathbf{R} + \mathbf{u}_\perp, \tag{9.20}$$

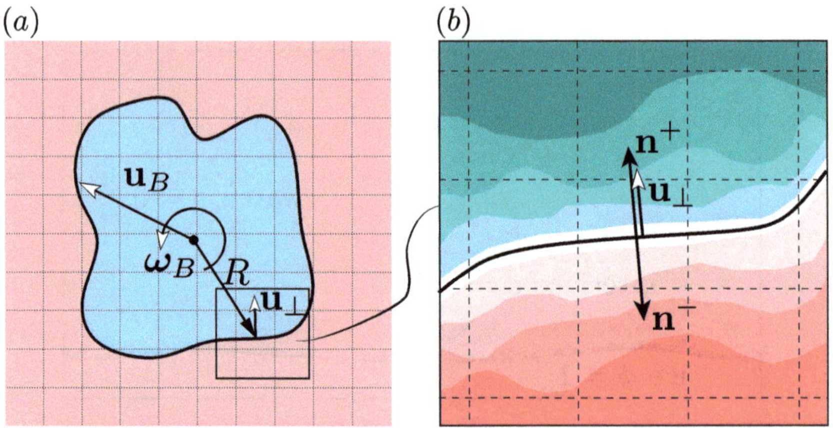

Figure 9.10 (a) Sketch of a solid object moving in a fluid with its surface receding because of melting. (b) Cartoon of the interface melting with the temperature variations across the interface.

while the boundary condition for the temperature field is simply $T = T_m$ since the interface is melting or freezing. Once again, these conditions can be imposed in the momentum and temperature equations via immersed boundary forcings as described in Chapter 4.

Additional difficulties in applying immersed boundary methods to melting problems are that solid bodies change shape and size during the evolution; thus their surface triangulation is distorted. The immediate consequence is that, even if the relative size of the Eulerian mesh and Lagrangian triangles is initially correct, this might not be the case at later times. Some of these issues have already been discussed in Chapter 7, since deforming bodies can also incur similar phenomena; for melting problems, however, the situation is exacerbated because if the solid melts completely, its volume disappears and even an initial perfect triangulation eventually yields degenerate elements with vanishing area or high skewness.

Usually, immersed boundary algorithms maintain the same triangulation in time or at least the same number of triangles if the body is deforming. For melting problems, this is not possible and the interface must be regularly triangulated to avoid integration instabilities or inaccuracies in the evaluation of local fluxes.

In Figure 9.11, we show the results for an ice sphere melting in liquid water obtained by a phase field method (Hester et al., 2020) and by an IBM with dynamic remeshing of the triangulated surface. Note that although in Figure 9.11b the ice volume is ≈ 15 percent of the initial value (Figure 9.11a), surface triangles are about the same size; without such a remeshing, the shrinking

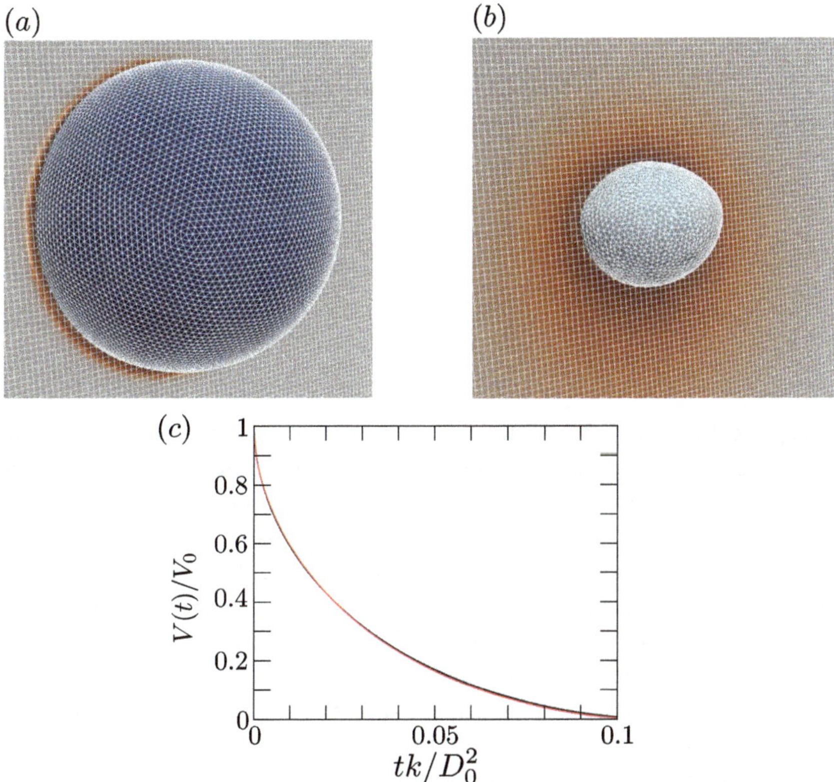

Figure 9.11 Ice sphere of initial diameter D_0 melting in liquid water at Rayleigh number $Ra_D = 10^3$, Prandtl number $Pr = 1.0$ and Stefan number $St = 1.0$: (a) initial condition; (b) configuration at $tk/D_0^2 = 0.06$ when the ice volume is ≈ 15 percent of the initial value V_0 (k is the thermal diffusivity). (c) Time evolution of the ice volume normalized by V_0 versus diffusive time: red solid line, IBM results with remeshing of the triangulated surface; black solid line, results obtained by the phase field method of Hester et al. (2020). (Courtesy of Kevin Zhong.)

size of triangle edges would destabilize the integration, requiring impractically small time steps and developing unphysical spikes and cusps.

While we can observe, from the comparison of Figure 9.11c, that the phase field method and IBM give identical results, we also note that the former method can be applied only when the solid body is convected by the fluid, while the latter approach has a general application. Indeed, this problem has been extensively validated and discussed by Zhong et al. (2025), where melting objects in homogeneous isotropic turbulence are simulated.

10

Numerical Examples

The present chapter is devoted to the application of immersed boundary methods to several numerical applications of increasing complexity. These range from two-dimensional flows around rigid bodies up to FSI three-dimensional flows.

The main aim of this series of examples is to make the reader familiar with the techniques and the algorithms; for this reason, a computer code is also made available so that the results for several of the presented cases can be replicated independently by the interested reader.

Finally, some additional examples are shown even if they cannot be immediately duplicated by running the provided code: They are intended as *challenges* for the interested researcher who wants to dare the implementation of new extensions of the method to venture into advanced applications of IBMs.

Before delving into the individual applications, we would like to provide a brief overview of the numerical code used to compute the flows. This will aid understanding of the algorithms and enable easy customization to meet varied needs.

10.1 Code Description

In this section, we present the implementation in a computer code of some of the IB methods that have been discussed throughout the text. The code is aimed at providing a quick start for the reader to get familiar with IBM for direct numerical simulations (DNSs), encompassing curved solid walls as well as fluid–structure interaction (FSI). For this reason, the distributed code IBbook is a compromise between simplicity and computational efficiency. Hence, on the one hand, IBbook is self-contained, and as it does not require the use of external libraries, the user needs only to copy–paste it to a laptop or a server with an available Fortran compiler; we recommend the GNU compiler `gfortran`. On the other hand, the whole code has been accelerated using openMP directives allowing the use of multiprocessor shared-memory computers for solving medium-sized three-dimensional flows.

The code can be retrieved at the following link.

`https://gitlab.com/vdv9265847/IBbookVdV/`

The directory `IBbookVDV` has several subfolders, the main one named `code` contains all the routines of the code. The latter is written in the programming language Fortran 77/90 and it can be compiled by typing *make* inside the same subdirectory. Details about the compiler and its options can be changed by editing the file *Makefile*. The other subfolders run* are the different test cases. Each test case has two input files, *ibbook.in* and *points.in*, which are explained shortly, and the geometry/geometries of the body/bodies in *.gts* format.

10.2 Governing Equations and Computational Domain

The fluid dynamics is solved by integrating numerically the three-dimensional, incompressible and unsteady Navier–Stokes equations

$$\frac{\partial \mathbf{u}}{\partial t} + \nabla \cdot (\mathbf{uu}) = -\nabla p + \frac{1}{Re}\nabla^2 \mathbf{u} + \mathbf{f}, \qquad \nabla \cdot \mathbf{u} = 0 \qquad (10.1)$$

in a Cartesian domain of size H_x, H_y, H_z, where x, y, z are the three Cartesian axes. The equations are made nondimensional by suitable velocity U and spatial scales L, thus yielding a temporal scale $T = L/U$. Accordingly, $\mathbf{u}$ and p are the nondimensional velocity and pressure fields and $Re = UL/\nu$ is the Reynolds number, with ν the kinematic viscosity of the fluid.

In this description, we use the standard italic font for continuum variables, while the corresponding variables in the numerical code are displayed in a monospaced font to help differentiate them clearly.

For instance, the three components of the velocity vector field $\mathbf{u} = (u_x, u_y, u_z)$ are stored in the three-dimensional arrays `q1,q2,q3`. The Navier–Stokes equations (10.1) are discretized using centered finite differences that are second-order accurate on a staggered mesh. In particular, `n1,n2,n3` grid points are used in the three directions and the corresponding grid values are stored in the one-dimensional arrays `x1c,x2c,x3c`, with the computational domain ranging from `x1c(1)` to `x1c(n1)` in the x-direction, and so on in the other two directions. The midpoints of the grid cells (ranging from 1 to `n1m = n1 - 1`

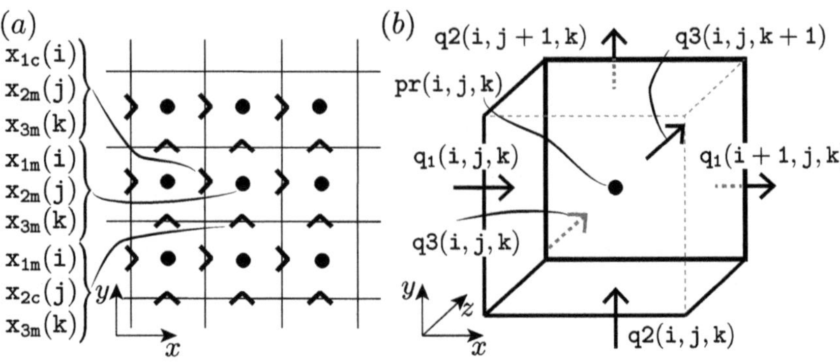

Figure 10.1 (a) Staggered grids in the x, y-plane and (b) corresponding collocation of the three-dimensional flow variables.

in the x-direction) are stored in x1m, x2m, x3m; see Figure 10.1(a). Hence, as shown in Figure 10.1(b), q_1 as well as all the terms of the momentum equation in the x-direction are located at the grid point x1c, x2m, x3m; similarly, q_2 and the terms of the relative equation are computed at x1m, x2c, x3m while q_3 and its terms are computed at x1m, x2m, x3c. Finally, the pressure pr and the equation of mass conservation are discretized in x1m, x2m, x3m (and the same would happen for any scalar quantity).

The code is biperiodic (*channel-type*) as the x- and y-directions have periodic conditions and uniform grid spacing in order to allow the use of FFT transforms when solving the Poisson equation, coming in the projection step of the fraction step method. On the other hand, Dirichlet or Neumann conditions can be applied in the z-direction and the distribution of nodes may be non-uniform. The boundary surfaces, corresponding to the minimum ($z =$ x3c(1)) and maximum ($z =$ x3c(n3)) z-coordinate, are denoted as *south* and *north* faces.

10.3 Input File

The input parameters to specify the size of the domain and the geometry of the immersed bodies are specified in ibbook.in and an example is reported as follows.

```
N1   N2   N3     NWRIT     NREAD    IRESET
241  121  241    1         0        0
ALX3     Z_TR       ISTR3
1.50     1.00       -1
ALX2     Y_TR
0.75     0.375
ALX1     X_TR
1.50     0.75
INSLWS     INSLWN   UINFLOW   COU
1          1        0.0       1.0
```

```
RE
5000
NTST          TMAX
10000000      11.
NSST IDTV DTMAX    DT       CFLMAX  CFLLIM  RESID
1     0   1.0e-3   1.0e-3   1.0     4.0     1.e06
TREST        TPRINT     TPIN          NSON
5.0e-01      1.0e-01    1.0e-02       2
IFSI  ONEDIR   STRONGCOUP
3     3        0
IFAD     IMLS      ALPHA_EXP_W      PLENGTH
0        1         0.7             2.5
N_BODIES BCBODY   GRAVITY
1        0        0.0 0.0 0.0
MLS BODY FILENAME        OPEN_POS     OPEN_NEG
triangulated.gts         1            0
BODY_DENS 2DYoungM K_BEND  K_VOL K_A_TOT K_A_LOC THICK CDAMP
1.00      1000.0   1.0d-4  0.0   0.0     0.0     0.01  0.0
AMPLITUDE    ANGULAR_FREQ    (imposed kinematics)
0.15         1.25
Kspring      Cspring    Ospring1    Ospring2    Ospring3
0.0          0.0        0.0         0.0         0.0
PosL1        PosL2      PosL3  VelL1  VelL2  VelL3
0.0          0.0        0.0    0.0    0.0    0.0
```

The input file thus consists of alternating lines: Odd-numbered lines are comments only, and serve as labels to indicate the meaning and order of the values on the following line. These lines are ignored by the parser. Even-numbered lines contain the actual input values that are read by the simulation.

n1, n2 and n3 are the numbers of grid points in the x-, y- and z-directions (note that n1, n2 and n3 cannot exceed m1, m2 and m3, defined in the file *param.f*, which statically allocates the dimensions of the arrays). n1 and n2 must be compatible with FFT (a combination such that n1, n2 = $2l \times 3m \times 5n + 1$, with l, m, n integers) while n3 (non-uniform direction) is free. NWRIT enables the writing of a restart file *restart.dat*, which allows the continuation of a simulation previously interrupted. To restart an existing simulation, the flag NREAD = 1 must be set, while if NREAD = 0 a new simulation is performed. When the IRESET is set to one, the counters (time, number of time steps, etc.) are reset when the simulation is restarted from a continuation file.

ALX3 is the length of the domain in the z-direction (H_z) and Z_TR translates the domain in the same direction, such that the grid nodes are distributed between [Z_TR, Z_TR+ALX3]. ISTR3 selects the distribution of the computational nodes in the z-direction: The grid is uniform if ISTR3 = -1, whereas the grid nodes could be clustered at the south wall (ISTR3 = 0), at the north wall (ISTR3 = 1) or at a different location (ISTR3 = 2). Instead, ISTR3 = 3 allows us to load the grid points from an external input file named *gridz.data*. ALX2 (ALX1) is the domain length in the y (x)-directions and Y_TR (X_TR) translates the grid nodes between [Y_TR, Y_TR + ALX2] ([X_TR, X_TR + ALX1]). For further details, see the subroutine *cordin*.

`inslws` and `inslwn` are, respectively, the flags to set the south and upper walls condition to Dirichlet (if the corresponding flag is 1) or free-slip (if the flag is 0). `uinflow` is the velocity value imposed at the south boundary (inlet) and `cou` is the convective velocity of the radiative condition imposed (through a Dirichlet condition) at the north boundary (outlet):

$$\frac{\partial \mathbf{u}}{\partial t} + \mathrm{cou}\nabla\mathbf{u} = \mathbf{0}, \tag{10.2}$$

which reduces to the no-slip condition when `cou` = 0.

`Re` is the Reynolds number of the flow.

`NTST` is the maximum number of time steps after which the run is interrupted and `TMAX` is the maximum integration time after which the simulation is stopped (even if this time is reached before `NTST` is exceeded)

`NSST` determines the time integration scheme: `NSST` = 1 selects a second-order Adams–Bashfort with a stability limit CFL < 1, `NSST` = 3 uses a third-order Runge–Kutta with a stability limit CFL < $\sqrt{3}$. `IDTV` sets the time integration strategy: If `IDTV` = 1, the time step size is computed dynamically by maintaining the stability limit CFL at the value `CFLMAX` (the resulting time step is anyway limited to the threshold value `DTMAX` in order to avoid accuracy issues). If `IDTV` = 0, the integration is performed with a fixed time step (`DT`) and the simulation is stopped if the CFL exceeds `CFLLIM`. `RESID` is the maximum allowed local divergence of the velocity field: The check is performed after every time step.

In the next line, `TREST` is the time interval for dumping three-dimensional snapshots of velocity and pressure fields, together with the restart file. `TPRINT` is the time interval for dumping two-dimensional flow slices, together with the instantaneous geometry of the immersed bodies. `TPIN` is the time interval for the printing of several diagnostics such as the probes sampled values, hydrodynamic loads on the immersed body and the position of its center of mass. `NSON` sets the number of numerical probes that sample the pointwise values of velocity components and pressure (see the subroutine *wson*). The positions of the probes are specified in the file *points.in*, which contains the list of the `NSON` coordinates of the probes.

`IFSI` switches between imposed body kinematics (including fixed bodies, when the flag is set to 0), FSI for rigid bodies (translation only) where the center of mass of the immersed body is governed by the Newton's equation (flag equal to 1) and FSI for deformable bodies (flag equal to 2). Furthermore, if `IFSI` = 3 the code solves for the body translation resulting from an imposed swimming motion in the body frame. In this case, equations (10.1) are solved in a reference frame translating with the body, thus accounting for the translation inertial forces. If ONEDIR is different from zero, the motion of the body center of mass is restricted to the coordinate direction indicated by ONEDIR itself (thus ranging between 1 and 3). The flag STRONGCOUP = 1 calls for strong coupling between the fluid and structure solvers; otherwise, loose coupling is used.

Direct Eulerian IBM is used when IFAD = 1, which is available in this implementation for fixed and closed bodies (i.e. those geometries whose surface encloses a volume). Otherwise, Lagrangian MLS IBM is used for IMLS = 1, with alphaW3 the coefficient of the exponential weight of the MLS (see Section 4.2.2). The offset distance in grid units from the wet surface of the probe used to compute the hydrodynamic loads is chosen as a compromise between having the probe as close as possible to the wet wall and having the corresponding MLS interpolation kernel (27 Eulerian points in three dimensions) on the same side with respect the wall, hence length_probe = 2.5.

n_bod is the number of immersed bodies and BCBODY sets the boundary condition on the body structure, which only applies for FSI cases (i.e. when IFSI $\geq$ 1). For BCBODY = 0 the body is unconstrained; otherwise, user-prescribed boundary conditions are applied, which are defined in the routines find_boundaries. The 3 × 1 array grav initializes body forces, such as the gravity force.

The following lines load the input data for the immersed bodies. The first entry of the array geofile contains the name of the immersed body (ending in *.gts*). iopen_pos and/or iopen_neg are turned to one if the body is wet in the positive and/or negative normal direction (please note that the normal direction is defined by the loaded geometry, as discussed in Section 3.1). The following line corresponds to the nondimensional body parameters, such as its mass density (rhos), elastic two-dimensional Young modulus (rke), bending stiffness (rkb) and additional constraints for volume (setting the input quantity rkv) and surface area (total and local through kat and kal). thickness is the shell thickness, which is needed to determine the material mass and bending rigidity, while the coefficient cvl sets the internal material dissipation. It should be noted that rhos is only used for FSI simulations (IFSI $\geq$ 1) and the other quantities for deformable bodies FSI (IFSI = 2).

The inputs Amp_IK and omeg_IK set the nondimensional amplitude and frequency oscillation of a harmonic motion in the y-direction imposed on the body when (IFSI = 0). Hence, the case of fixed rigid bodies is set using IFSI = 0 and Amp_IK = 0.

In the case of FSI on rigid bodies (IFSI $\geq$ 1, translation without rotation has been considered), Kspring and Cspring are the spring stiffness and dashpot dissipation applied to the center of mass of the object, where the array Ospring contains the spring equilibrium position.

Finally, posL_n and velL_n set the initial condition on the nondimensional position and velocity of the center of mass of the body.

For n_bod > 1, the input parameters for the other bodies are appended in *ibbook.in* following the same scheme.

10.4 Parameters and Global Arrays

All parameters and common arrays are defined in the file *param.f*, which is included by most of the subroutines. In addition to the input parameters already defined, some relevant quantities are as follows.

- m1, m2, m3 are the maximum number of computational nodes of the fluid grid. Hence, all volume arrays are dimensioned as q1(m1, m2, m3).
- rhs, qcap, dph, dq3 are temporary volume arrays which are needed for the solution of the discrete Navier–Stokes equations.
- amph(m3), acph(m3), apph(m3) are the tridiagonal matrix coefficients of the elliptic Poisson equation.
- dx1 and dx2 are the inverse of the (uniform) grid size in the x- and y-directions (n1m/alx1 and n2m/alx2). In the non-uniform z-direction, dx3 is defined as dx3 = n3m, and the inverse of the local grid spacing is then obtained using the grid metrics terms g3c as udx3 = dx3/g3c(kc), where kc is the index of the local z-coordinate.
- imaxv(3), jmaxv(3), kmaxv(3) and iminv(3), jminv(3), kminv(3) are the indices of the node corresponding to the maximum and minimum velocity in the domain. The three entries correspond to the three velocity components.
- The real array coson(nsonmax, 3) contains the coordinates of the probe locations (read from *points.in*) and the integer array mason(nsonmax, 3) gives the corresponding grid indices. nsonmax is the maximum number of probes.
- gam, rom, alm store the coefficients of the time marching scheme for the explicit terms, either Adams–Bashforth or third-order Runge–Kutta.

- `qb1s(m1,m2)` and `qb1n(m1,m2)` are the *x*-velocity fields at the south and north walls. The other two components are stored in `qb2s(m1,m2)`, `qb2n(m1,m2)` and `qb3s(m1,m2),qb3n(m1,m2)`. The corresponding `dqb*s(m1,m2), dqb*n(m1,m2)`, with $* = 1, 2, 3$, are the velocity increments at the south or north boundaries within each time step.
- `visc(m3)` is the nondimensional flow viscosity which is equal to $1/Re$, but it may be artificially increased when approaching the boundaries to create a viscous sponge region to avoid spurious wave reflections in open flow simulations.
- `mpun` is the maximum number of first external (and internal) grid points used to apply the Eulerian IBM. `forclo_q1,q2,q3` are volume arrays which are 1 everywhere and 0 in the locations where the Eulerian IBM is imposed. They are needed to preserve the enforcement of the IBM force during the solution of the provisional velocity. Three different arrays are needed because the grid is staggered. `indgeofx(4,mpun,3)`, `indgeoefx(4,mpun,3)` and `indgeoifx(4,mpun,3)` contain the list of the first external, second external and internal grid points. The first index switches between the velocities and pressure staggered grids, the second one lists the different forcing points, and the last one corresponds to the three node indices (i, j, k). `distbfx(4,mpun)` is the interpolation weight to determine the velocity at the first external grid point from the second external one (and assuming fixed body).
- `for_q1,2,3` are the volume arrays of the MLS-IBM (already transferred to the Eulerian grid) and `nex = 27` is the number of Eulerian points used in the MLS support domain.
- `nv(n_bod),ne(n_bod),nf(n_bod)` are arrays containing the number of vertices, edges and faces of each immersed body. The corresponding total numbers are stored in the scalars `nvtot,nftot,netot`. `count2` is the number of body vertices where the boundary condition should be enforced and `boundary2` collects the corresponding node indices. `ntri` is the number of triangles where the IB-MLS is enforced, which is equal to `nftot` in the cases considered here.
- The mesh connectivity of the triangulated geometries is expanded in several arrays. In particular, `n_edge_of_vert` lists how many edges are passing through each vertex. Hence, the corresponding `edge_of_vert` and `vert_of_vert` are made by the indices of the edges passing through each vertex and by the vertices connected to it. `vert_of_edge` contains the indices of the two vertices of any edge, `face_of_edge` stores the indices of the faces sharing a given edge (except for naked edges having only one face, each edge is shared by two faces). The array `v1234` gathers the

indices of the four vertices of the two faces sharing the same (non-naked) edge. `vert_of_face` and `edge_of_face` have the indices of the three vertices and three edges of each face.

- `LabelV(nvtot)`, `LabelE(netot)`, and `LabelF(nftot)` are one-dimensional arrays indicating to which body each vertex, edge or face belongs.

- The array `iTE` collects the indices of three trailing-edge vertices of the body; this is needed for the demo of the flapping flag.

- `xyz0(3, nvtot)` and `xyz(3, nvtot)` list the initial (free-stress) and instantaneous vertices coordinates. `xyzv` and `xyza` are the corresponding velocity and accelerations. Additionally, the arrays `xyzm1`, `xyzvm1`, `xyzam1` store the body configuration and dynamics at the previous time step.

- `dist0(netot)` and `dist(netot)` are the initial and instantaneous edge lengths, `theta0(netot)` and `theta(netot)` are the corresponding angles between the two faces sharing the same edge. Similarly, `sur0(nftot)` and `sur(nftot)` are scalar arrays with the local area of each triangular face, whereas `Surface0(n_bod)`, `Surface(n_bod)` and `Volume0(n_bod)`, `Volume(n_bod)` are the surface and volume of each immersed body.

- `tri_ver(nftot, 9)` contains the nine coordinates of the three vertices of a face. `tri_bar(nftot, 3)` has the three coordinates of the center of each face, whereas the components of the corresponding normal vectors are given in `tri_nor(nftot, 3)`. `vel_tri(nftot, 3)` and `acc_tri(nftot, 3)` are the components of the velocity and acceleration of the center of each triangle.

- `fxyze(3, nvtot)` and `fxyzi(3, nvtot)` are the external (hydrodynamics) and internal (elastic) forces applied on each vertex. `mass_of_vert` is the mass associated to each vertex.

- For each of the four staggered grids (u_x, u_y, u_z and p), `tri_cell(nftot, 3, 4)` stores the three indices of the Eulerian cell containing the center of any triangular face. `tri_cell_probe(nftot, 3, 4, 2)` does the same for the corresponding positive and negative probes.

- When solving the dynamics of the center of mass of a body, one needs to compute the resulting pressure `For_pres(n_bod, 3)` and viscous `For_visc(n_bod, 3)` force, along with the one of the body forces `For_grav(n_bod, 3)`.

- In the strong coupling (`IFSI = 2` and `STRONGCOUP = 1`), `n_max_iter` and `err_max` are the maximum number and maximum error admitted in the corrector iterations. `undr` is the under-relaxation coefficient for the corrector.

10.5 Code Routines

IBbookVDV/code/main.f

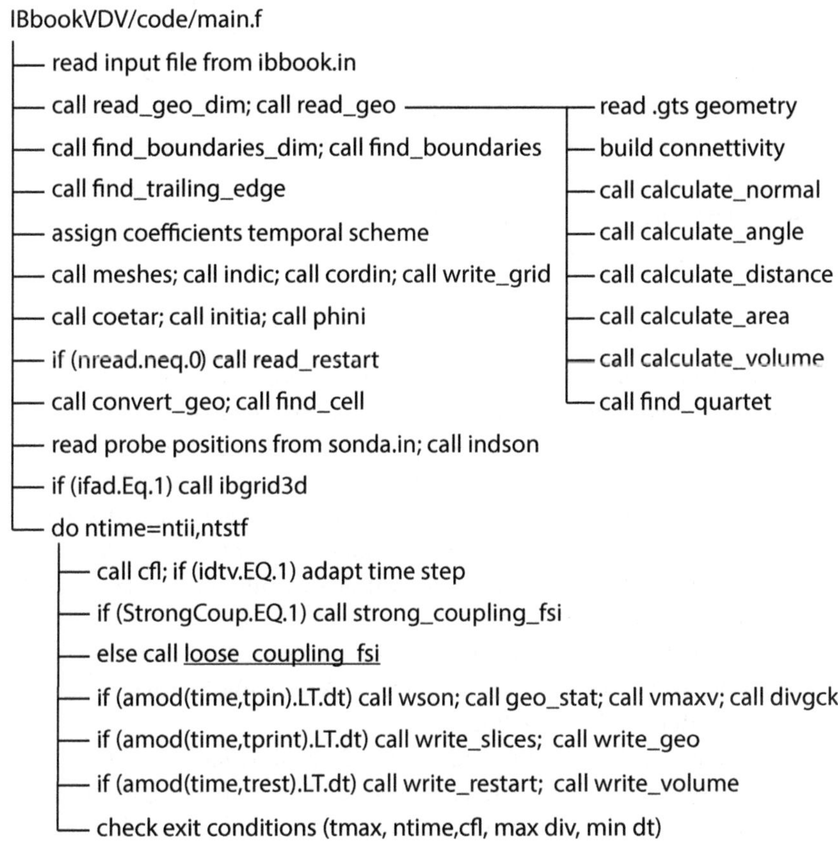

Figure 10.2 Outline of the code.

Figures 10.2 and 10.3 give an outline of the code, where simple text (such as *read input . . .*) briefly describes what operation is done, while the *call* statement indicates that a routine is invoked.

The main program is *IBbookVDV* (located in *code/main.f*), which reads the input file *ibbook.in* explained in Section 10.3. *read_geo_dim* reads the number of vertices, edges and faces nv, ne, nf of each immersed body, then *read_geo* reads all the geometries, appending their coordinates in a single array xyz. The same is done for the other structural arrays, including the mesh connectivity (e.g. $edge\_of\_vert$, $face\_of\_edge$). The same routine also computes the face-normal vectors (*calculate_normal*) and it computes the initial angle $theta0$ (*calculate_angle*), edges length $dist0$ (*calculate_distance*), body

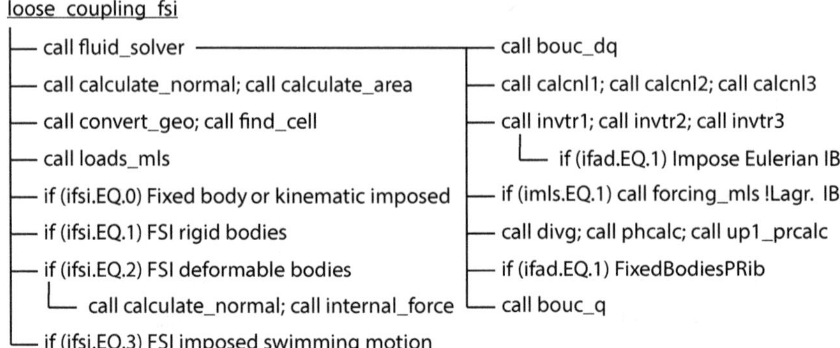

Figure 10.3 Outline of the loose-coupling routine.

$\mathtt{Surface0}$ and triangle area $\mathtt{sur0}$ (*calculate_area*), body volume $\mathtt{Volume0}$ (*calculate_volume*) and the quadruplets $\mathtt{v1234}$ (*find_quartet*).

If $\mathtt{BCBODY} \geq 1$, the routines *find_boundaries_dim* and *find_boundaries* build the list $\mathtt{boundary2}$ of $\mathtt{count2}$ mesh vertices to keep hinged or blocked during the simulation. The routine *find_trailing_edge* determines the indices of three points (maximum, average and minimum x-coordinate) on the trailing edge of the body. The latter is determined analytically for the demo of the flapping flag (see Section 10.13).

Then, depending on the computational flag $\mathtt{nsst}$ the time marching coefficients for the explicit terms are either set to an Adams–Bashforth scheme or to a third-order Runge–Kutta method.

The routine *meshes* computes $\mathtt{dx1}, \mathtt{dx2}, \mathtt{dx3}$ while *indices* computes the stencil indices for the finite-difference method, also accounting for the presence of the Dirichlet or Neumann condition in the z-direction and of the x, y periodicity. *cordin* creates the uniform $\mathtt{x1c}, \mathtt{x1m}, \mathtt{x2c}, \mathtt{x2m}$ and non-uniform $\mathtt{x3c}, \mathtt{x3m}$ staggered grids, as well as the metric of the grid in the z-direction $\mathtt{g3m}, \mathtt{g3c}$. *write_grid* writes the grid file *grid.xyz*: As in the IBM the Eulerian grid is kept fixed, this file is saved only once before starting the time marching of the solution.

The finite-difference coefficients on the non-uniform z-grid are determined in *coetar*, while the volume arrays are initialized *initia*. In particular,

$\mathtt{q3(:,:,:)} = \mathtt{uinflow}, \mathtt{qb3s(:,:)} = \mathtt{qb3n(:,:)} = \mathtt{uinflow}$. The coefficients of the Poisson equation for the pseudopressure ($\mathtt{dph}$) are calculated in *phini*.

If $\mathtt{nread} \neq 0$, the simulation is restarted from an existing one by loading the restart file *restart.dat* containing the initial conditions for the fluid and the structures.

convert_geo computes the arrays `tri_ver,vel_tri,acc_tri,` `tri_bar` out of `xyz,xyzv,xyza`. Moreover, it computes the bounding-box of the body, which will be used to speed up *find_cell* to find the Eulerian cell containing each Lagrangian marker (triangle center) as well as the corresponding probes.

The coordinates of the probe positions are read from *points.in* and *indson* determine the indices of the closest Eulerian grid point for each probe.

If `IFAD.EQ.1` (direct Eulerian IBM is active), then the ray-tracing algorithm (*ibgrid3d*) identifies the first and second external grid points, the corresponding interpolation coefficients and the internal grid points (closed surface bodies).

The governing equations are then time marched. Firstly, *cfl* computes the maximum CFL in the domain (normalized by the time step `cflm = CFL/dt`) and it adapts the time step if $IDTV = 1$.

If `STRONGCOUP.EQ.1`, an iterative predictor–corrector scheme for the strong coupling FSI is run, while a loose-coupling step is carried out otherwise. Afterward, every `tp` in the velocity and pressure at the probe locations (*wson*), flow statistics *geo_stat*, maximum velocity (*vmaxv*), velocity divergence (*divgck*), are computed and possibly stored to file (see Section 10.6). Moreover, two-dimensional slices of the flow field and the instantaneous body configuration are saved every `tprint`. The restart file is updated using *write_restart*, and the volume data are dumped to file using *write_volume*.

The temporal loop keeps running until one of the exit criteria is met:

- simulation time larger than the final time: `time > tmax`
- step number larger than maximum allowed: `ntime > ntst`
- time step smaller than minimum allowed: `dt < dtmax/5`
- unsatisfied CFL condition: `cflm > cfllim`
- unsatisfied condition on the mass conservation: $\nabla \cdot \mathbf{u} =$ `dmax > resid`.

Let us now turn our attention to the FSI routine *loose_coupling_fsi*. Firstly, the fluid mechanics is solved by calling *fluid_solver*, which closely follows the fractional step method detailed in Section 4.4.1. In particular, the variation of the velocity fields at the south or north boundaries (`dqb1s,dqb1n,dqb2s` and so on) is updated in *bouc_dq*. The nonlinear terms of the three components of the momentum equations are assembled in *calcnl1,2,3*, and the three components of the provisional velocity are computed in *invtr1,2,3* (solving tridiagonal systems). Within these routines, the no-slip condition is enforced by Eulerian IBM if $IFAD = 1$. Conversely, if $IMLS = 1$ the IBM-MLS forcing is computed and applied in *forcing_mls* to the provisional velocity. In order to enforce mass conservation, the divergence field of the provisional velocity is computed (*divg*), the corresponding Poisson equation is solved (*phcalc*) and, lastly, the velocity is projected in the solenoidal space and pressure is updated in *up1_prcalc*. Then, in the case of Eulerian IBM (which is here implemented for closed and fixed bodies), the velocity in the internal grid points can be set to zero (as discussed in Section 4.3) and, as a consequence, the pressure can be set as well to a constant value in *FixedBodiesPRib*. Lastly, the velocity field at the north or south boundaries is updated (*bouc_q*).

Coming back to *loose_coupling_fsi*, at each time step the body geometrical quantities are updated by calling a set of routines already encountered (*calculate_normal*, *calculate_area*, *convert_geo* and *find_cell*). Then in *load_mls* the hydrodynamic loads as well as the resulting (pressure and viscous) force applied from the fluid on the immersed bodies are computed. Note that also when Eulerian IBM is adopted, the hydrodynamic loads are computed by using the MLS kernel introduced in Section 6.3.

If the flag $\mathtt{IFSI} = 0$, then a harmonic motion in the transversal y-direction is applied to the body, whose amplitude and angular frequency are set in *ib-book.in*. If the oscillation amplitude is set to zero, we retrieve the case of flow around a fixed body.

On the other hand, if $\mathtt{IFSI} = 1$ the FSI for rigid bodies is solved (translation only). Hence, the position of the center of mass of the body is governed by Newton's equation, which balances inertia, the resulting hydrodynamic force, body forces as well as elastic and dissipation forces in the case where a spring or dashpot is applied to the body (we recall that the spring constant and dissipation coefficient are given in the input file).

The flag $\mathtt{IFSI} = 2$ corresponds to full FSI for deformable bodies, where the internal passive stresses are computed in *internal_force* based on the spring network approach introduced in Section 7.4.2.

As a last case, when $\mathtt{IFSI} = 3$ a swimming motion is applied to the immersed body. The Navier–Stokes equations are solved in a body frame translating with the center of mass of the body, but always keeping the same orientation to the laboratory frame. In this way, only inertial translational forces enter the body frame dynamics. The dynamics of the center of mass of the swimmer is then solved by balancing inertia and the resulting hydrodynamic force.

10.6 Outputs

When the code is run, apart from some initial headings, it displays at every iteration the number of time steps ($\mathtt{ntime}$), the nondimensional integration time and the time step size (which may vary if `idtv` = 1). Moreover wall clock time per time step and the maximum velocity divergence in the flow field are printed as follows.

```
Time step:              5497 / Time:   0.10992E+02 / Dt:   0.20000E-02
WTIME LOOP   :      0.634000000000000001
Max Div      :      0.57376E-12
```

Every `tpin` also the minimum and maximum velocity in the three Cartesian directions are printed, along with the corresponding node indices. An example is given in the following.

```
MinMax X-vel:    -0.61314E+00     42     65    163     0.40276E+00     44
        68   165
MinMax Y-vel:    -0.39708E+00    157     45      1     0.28993E+01     41
        67   163
MinMax Z-vel:    -0.82249E+00     41     67    163     0.47505E+00     41
        65   163
```

The files *p***.out* contain the measurements of the numerical probes. The output of the first probe (whose coordinates are in the first line in the input file *points.in*) is in the file *p001.out*. Thus *p002.out* contains the output of the second probe and so on. These files have eight columns; the first one is time, then there are the three velocity components and pressure. The last three quantities are the three vorticity components. Everything can be computed and sampled by changing the routine *wson*. The latter also prints (every `tpin`) *kenergy.out*, where the first column is time and the second one is the total kinetic energy of the flow $0.5 \int_\Omega \mathbf{u} \cdot \mathbf{u} d\Omega$, with Ω the Cartesian flow domain, which in Fortran style becomes the following.

```
    energ = 0.
      do k=1,n3m
        do j=1,n2m
          do i=1,n1m
            energ = energ + 0.5*(q1(i,j,k)**2+
 &                 q2(i,j,k)**2 + q3(i,j,k)**2)*
 &                 g3m(k)/(dx1*dx2*dx3)
          end do
        end do
      end do
```

Still every `tpin`, the routine *geo_stat* appends to the file *dyna00001.out* some quantities related to the dynamics of the first immersed body: time, the three components of the resulting pressure force, the three components of the resulting viscous force, the center of mass position, velocity and acceleration of the y and z components.

```
        do mm=1,n_bod
        write(ipfi,99) mm
        namfi='dyna'//ipfi//'.out'
        open(197,file=namfi,status='unknown',position="append")
```

```
      write(197,546) time,
&            For_pres(mm,1),For_pres(mm,2),For_pres(mm,3),
&            For_visc(mm,1),For_visc(mm,2),For_visc(mm,3),
&            posL_n(mm,2),velL_n(mm,2),accL_n(mm,2),
&            posL_n(mm,3),velL_n(mm,3),accL_n(mm,3)
      close(197)
      enddo
```

A different file is saved for each mm body. Of course, the routine can be customized, for instance also printing the x-displacement if needed. The same routine also dumps a dedicated output file for the flag test case *TE_flag.out* (see section 10.13), where the three-dimensional displacement of three points on the trailing edge is collected.

```
open(198,file=namfi,status='unknown',position="append")
      write(198,546) time,
&              xyz(1,iTE(1)),xyz(2,iTE(1)),xyz(3,iTE(1)),
&              xyz(1,iTE(2)),xyz(2,iTE(2)),xyz(3,iTE(2)),
&              xyz(1,iTE(3)),xyz(2,iTE(3)),xyz(3,iTE(3))
      close(198)
```

Every `tprint` the body geometry is saved in a file like *struc_01_00000050.vtk*, where the first number indicates the body number and the last one is the simulation time scaled by a factor (typically 10^3 in order to get an integer). The data file can be customized (modifying the routine *write_geo*) in order to save any physical quantities defined on the body mesh, such as the pressure loads.

```
    write(11,*)''
    write(11,*)'POINT_DATA ',nv
    write(11,*)'Scalars ext_loads FLOAT'
    write(11,*)'LOOKUP_TABLE default'
    do i=1,nv
     write(11,*)  sqrt(fxyzep(1,i)**2+fxyzep(2,i)**2+fxyzep(3,i)
        **2)/sur_nod(i)
    end do
```

The string *POINT DATA* indicates that this scalar field is defined on the vertices of the body mesh. Nevertheless, cell-based quantities also could be dumped, as follows.

```
        write(11,*)''
        write(11,*)'CELL_DATA ',nf
        write(11,*)'Scalars c_data1 FLOAT'
        write(11,*)'LOOKUP_TABLE default'
        do i=1,nf
           write(11,*) c_data1(i)
        end do
```

With the same time interval `tprint`, two-dimensional flow slices of the x–z- and y–z-planes are saved in files like *fieldXZ_00000050.dat* and *fieldYZ_000000 50.dat*, where the number indicates the simulation time (as for the body file). The slices contain the three components of velocity and pressure. The printing routine *write_slices* can be modified to save additional fields and/or save additional planes or modify the predefined positions.

As anticipated earlier, since the Eulerian grid of the fluid dynamics is not changing during the simulation the grid file *grid.xyz* is stored only once at the beginning of the simulation by the routine *write_grid*. Then, every `trest` the volume flow field is saved in a file like *volume_00000050.q* (which is equivalent to the *plot3d* format) where again the number in the string indicates the time. In addition to the three velocity components and pressure, the data files contain a scalar field marking the cells where the IBM force is applied.

All these data files (geometries, flow slices and flow volumes) can be used for visualization using the software Paraview. With minimal amendments, they can be made compatible with other visualization software. Figure 10.4 shows an example of the visualization of the output files. The two orthogonal planes correspond to the *fieldXZ_*.dat* and *fieldYZ_*.dat* files, while the black edges of the Cartesian domain indicate the volume data field within *volume*q*. The swimming manta ray has been loaded through the *struc_01_*.vtk* file.

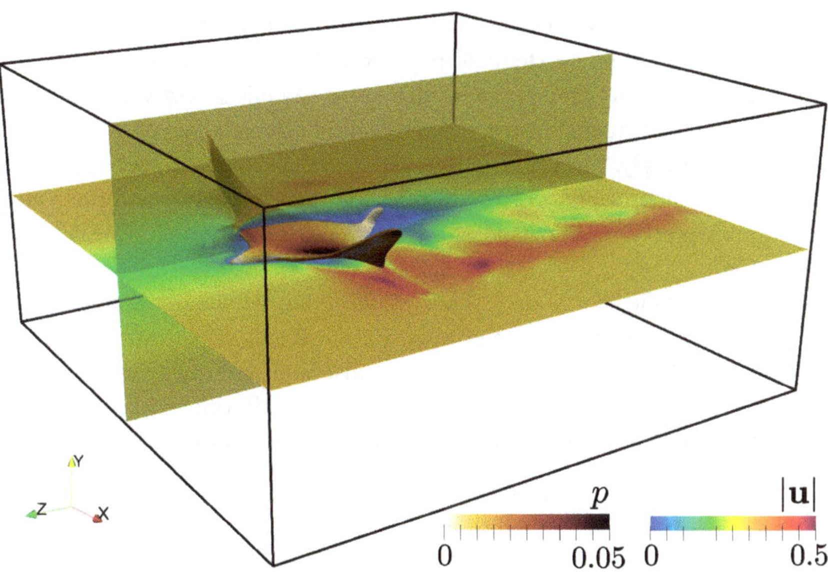

Figure 10.4 Example of visualization of the structural, two-dimensional and three-dimensional output files.

In the rest of this chapter, we will describe some specific applications that are provided with representative results that can be used as benchmarks. Before introducing the specific immersed boundary examples, it is useful to point out that when all the flags activating IBMs are turned off, the provided code is

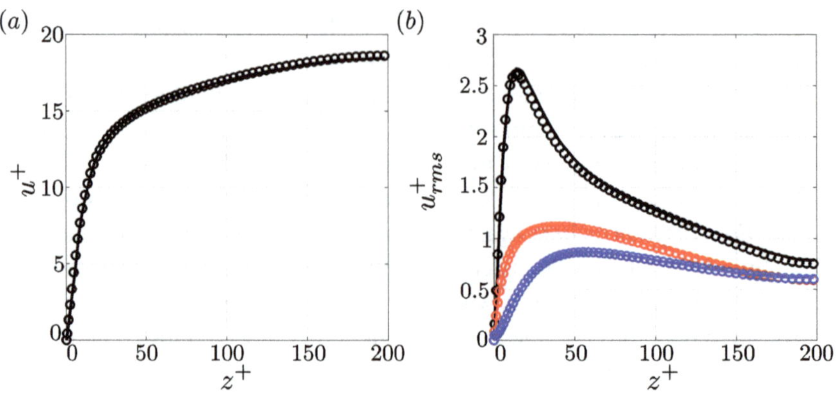

Figure 10.5 Mean (a) and rms (b) velocity profiles (streamwise, spanwise and wall-normal components in black, red and blue), plotted in viscous units, for the turbulent channel flow at $Re_\tau = 200$. The solid lines are the results obtained with the present code, symbols are the benchmark results by Quadrio et al. (2016).

just a standard flow solver which integrates the incompressible Navier–Stokes equations in a three-dimensional biperiodic Cartesian domain.

This configuration is extremely popular within the CFD community as it represents the setup of the turbulent channel, a canonical flow which has been serving as a training and benchmark problem for generations of fluid dynamicists. Accordingly, the interested reader might try to reproduce the results for this flow, and in Figure 10.5 we report the mean velocity and rms velocity profiles plotted in the standard viscous units (see the following for their definition).

For this flow the "north" and "south" walls of the computational domain have to be set as no-slip; this is achieved by `cou = 0` and `uinflow = 0` and the box, of size $H_x = 4\pi$, $H_y = 2\pi$, $H_z = 2$, must be discretized, respectively, by `n1 = 257`, `n2 = 257`, `n3 = 129` nodes. These are uniformly spaced except for the z-direction where a Chebyshev distribution has been used

$$z(k) = \cos\left(\frac{\pi(k - 0.5)}{\texttt{n3}}\right), \quad \text{for} \quad k = 1, \ldots, \texttt{n3},$$

(which can be easily added to the already available node distributions in the subroutine *cordin*).

The second (easy) change needed is the forcing term for the flow, which can be added in the form of an additional, constant pressure gradient: We have added a positive unity constant to the existing pressure gradient in the x-momentum equation (subroutine *calcnll*) which, with the input value `Re = 200`, produces the results of Figure 10.5.

The scaling in non-dimensional viscous units is obtained by computing the friction velocity u_τ and viscous length δ^+ via

$$u_\tau = \sqrt{\frac{1}{Re}\left\langle \left.\frac{\partial u_x}{\partial z}\right|_W \right\rangle}, \qquad \delta^+ = \frac{1}{u_\tau Re},$$

where $\langle \ldots \rangle$ is the surface and time average of the wall-normal derivative of the streamwise velocity evaluated at the upper/lower wall. The variables in Figure 10.5 are $u^+ = u/u_\tau$ and $z^+ = z/\delta^+$.

10.7 Two-Dimensional Flow Around Two Cylinders in Tandem

The problem proposed in this section was investigated by Borazjani and Sotiropoulos (2009) and successively reconsidered by Geraci et al. (2017) for uncertainty quantification analysis. The main features of the computational setup are summarized in Figure 10.6 with two rigid cylinders exposed to a two-dimensional, horizontal, uniform flow of velocity magnitude U. The cylinders, of diameter D, are aligned along the unperturbed velocity and their centers are at a distance L_x. The bodies are mounted elastically on identical spring/absorber assemblies of elastic constant K and damping coefficient c and their motion is restricted to vertical oscillations so that each body has only a single degree of freedom.

The computational domain is $H_x \times H_y$ with $-8D \leq x \leq 24D$ and $-8D \leq y \leq 8D$ with the flow coming from left to right in the horizontal direction. The tandem arrangement is such that $L_x/D = 1.5$ with the front cylinder center located at $(0, 0)$ and that of the rear counterpart at $(1.5, 0)$.

A non-uniform grid of 379×454 nodes is used, with a uniform grid spacing of $0.02D$ in the vicinity of the cylinder. The Lagrangian markers on the surface of the cylinders are distributed uniformly, with a spacing of $0.014D$, which is equal to 0.7 times the local Eulerian grid size. A constant time step is used, $\Delta t = 0.002D/U$, resulting in a $CFL = 0.4$. Inlet and outlet boundary conditions are imposed on the left and right boundaries, whereas free-shear wall conditions are imposed for the horizontal ones.

The flow is governed by the Navier–Stokes equations for an incompressible, viscous fluid which, in nondimensional form, read

$$\frac{\partial \mathbf{u}}{\partial t} + \mathbf{u} \cdot \nabla \mathbf{u} = -\nabla p + \frac{1}{Re}\nabla^2 \mathbf{u} + \mathbf{f}, \qquad \nabla \cdot \mathbf{u} = 0, \tag{10.3}$$

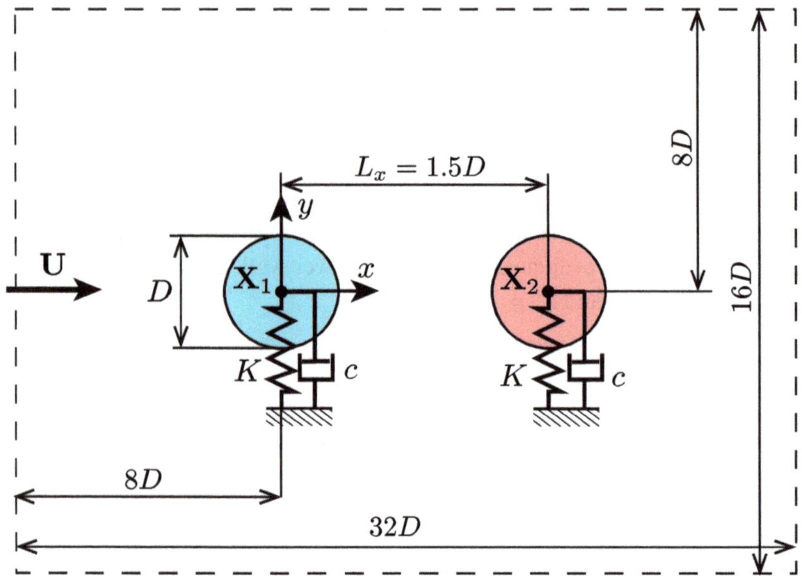

Figure 10.6 Schematic of the computational setup for the two cylinders in tandem in a uniform flow.

with $Re = UD/\nu$ the Reynolds number obtained by using U and D, respectively, as velocity and length scales.

The position of each cylinder centroid $\mathbf{X}_i$ is given by the equation

$$M\frac{d^2\mathbf{X}_i}{dt^2} + c\frac{d\mathbf{X}_i}{dt} + K\left(\mathbf{X}_i - \mathbf{X}_i^0\right) = \mathbf{F}_i(t), \qquad i = 1, 2 \tag{10.4}$$

with $\mathbf{X}_i^0$ the spring neutral position coinciding with the initial location of the cylinder, M its mass and $\mathbf{F}_i(t)$ the hydrodynamic loads produced by pressure and viscous forces. Since the cylinders are allowed to oscillate only in the vertical direction, equation (10.4) is used to determine only the vertical displacement Y_i.

All the cases presented in this section are obtained for a damping coefficient $c = 0$, therefore the natural frequency of the oscillations is given by

$$f = \frac{1}{2\pi}\sqrt{\frac{K}{M}}, \tag{10.5}$$

which is used to define the nondimensional reduced velocity

$$U_{\text{red}} = \frac{U}{fD}. \tag{10.6}$$

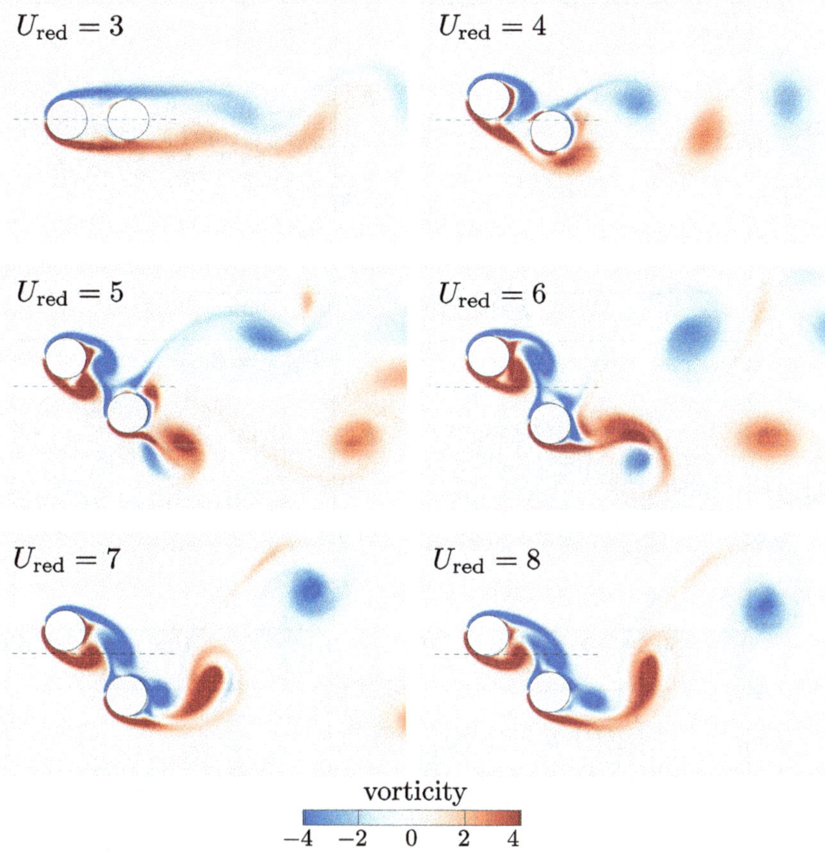

Figure 10.7 Instantaneous vorticity contours around two tandem cylinders, each with one degree of freedom for transverse vibrations (1-DOF case), at $Re = 200$, $M_{\mathrm{red}} = 2$ at increasing values of U_{red}. The horizontal dashed line indicates the rest position of the cylinders. The reported time instants correspond to those at which the cylinders are at their maximum distance.

This, together with the reduced mass $M_{\mathrm{red}} = M/(\rho_f D^2)$, completely defines the problem in terms of nondimensional parameters, with ρ_f the density of the fluid.

The Reynolds number of all the simulations has been kept fixed to $Re = 200$ while U_{red} has been varied from 3 to 8 with six equispaced values.

Representative results are given in the figures here, and they show instantaneous snapshots of the vorticity field for different values of U_{red} (Figure 10.7), time histories of the vertical displacement of the front and rear cylinders (Figure 10.8) and their maxima (Figure 10.9).

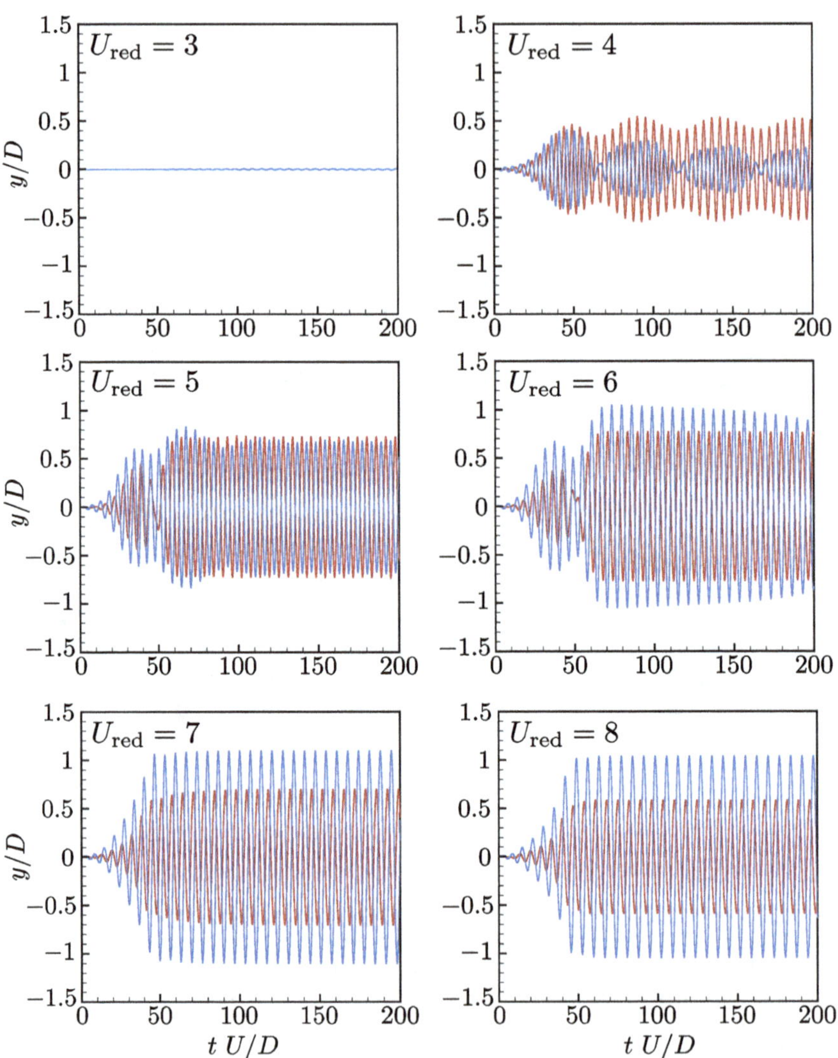

Figure 10.8   Time history of cross-flow displacement (y/D) for 1-DOF two tandem cylinders at $Re = 200$, $M_{\mathrm{red}} = 2$ at increasing values of U_{red}. Red lines: front cylinder; blue lines: rear cylinder.

10.8   Sedimentation of a Two-Dimensional Elliptic Particle

The present flow is that generated by a two-dimensional elliptic particle of axes a and b sedimenting under a uniform gravity field $\mathbf{g}$ in a stagnant fluid.

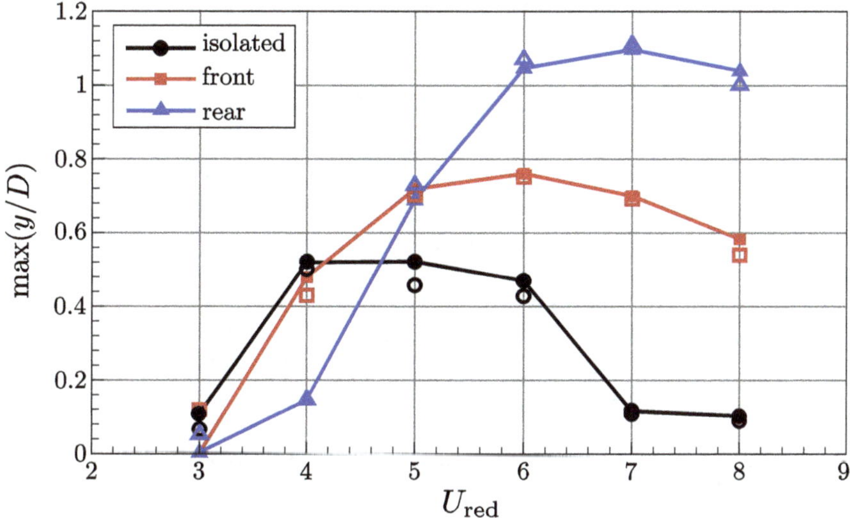

Figure 10.9 Maximum vertical displacement versus reduced velocity for 1-DOF two tandem cylinders at $Re = 200$, $M_{\mathrm{red}} = 2$, compared with an isolated cylinder. Open symbols indicate results by Borazjani and Sotiropoulos (2009).

The problem has already been considered by Xia et al. (2009), who conducted numerical experiments using a multiblock lattice Boltzmann method as well as a finite-element technique, and by de Tullio and Pascazio (2016), who used those results to validate their IBM procedure.

A schematic of the computational setup is given in Figure 10.10: The main scaling length is the major axis of the ellipse a and its aspect ratio is $b/a = 0.5$. The computational domain is $H_x = 4a$ in the horizontal dimension and $H_y = 28a$ in the vertical dimension, with the gravity vector $\mathbf{g}$, which is antiparallel to $\mathbf{y}$. All the boundaries of the computational domain are no-slip.

The particle is denser than the fluid, with $\rho_s/\rho_f = 1.1$ being the subscripts s and f used for the solid and the fluid, respectively. Placing the origin of the axes in the lower left corner of the domain, the initial position of the particle centroid is $(2a, 24a)$ and its orientation is $\theta = 45°$.

Let V_t be the terminal settling velocity of the particle, the Reynolds number of the flow is $Re = V_t a/\nu = 12.5$, with ν the kinematic viscosity of the fluid, and $Fr = V_t/\sqrt{ga} = 0.126$ the Froude number.

de Tullio and Pascazio (2016) used uniform Cartesian grids with $0.005 \leq \Delta x = \Delta y \leq 0.05$ showing that, for the present parameters, $\Delta x \leq 0.02$, corresponding to a mesh of 201×1401 nodes, yielded grid-independent results.

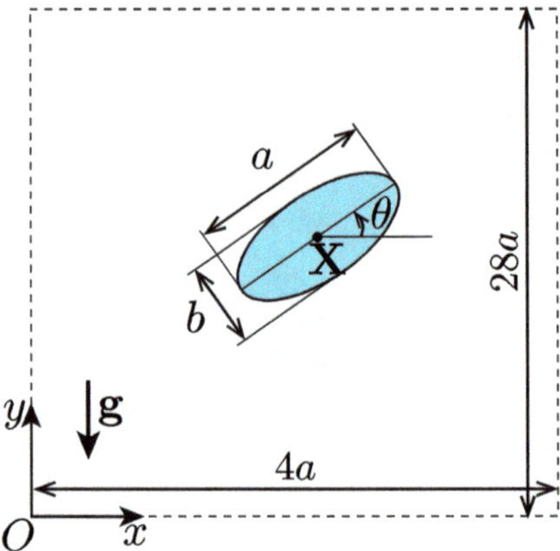

Figure 10.10 Sketch (not to scale) of the computational setup for the sedimentation of an elliptic particle in a stagnant fluid.

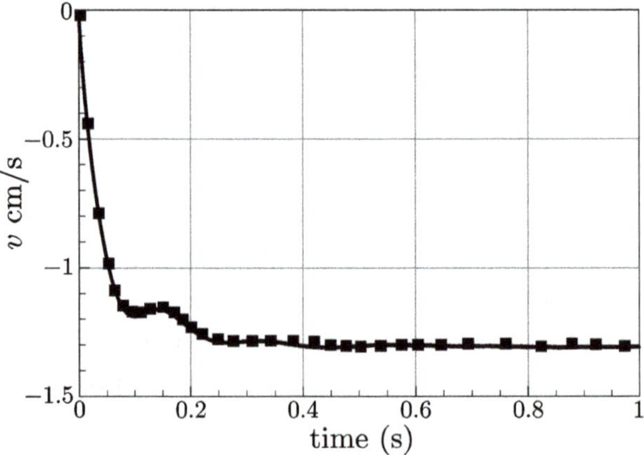

Figure 10.11 Sedimentation velocity for the elliptic particle sedimenting in a confined channel with $Re_t = 12.5$, $Fr_t = 0.126$. The present results (lines) are compared with finite-element numerical results of Xia et al. (2009) (symbols).

Lagrangian markers are uniformly spaced along the ellipse boundary with a stride $0.77\Delta x$.

Figure 10.11 shows the time evolutions of vertical velocity $v(t)$, while horizontal position of the centroid x and angular orientation θ are shown in figure 10.12, demonstrating a very good agreement with the finite-elements numerical data by Xia et al. (2009).

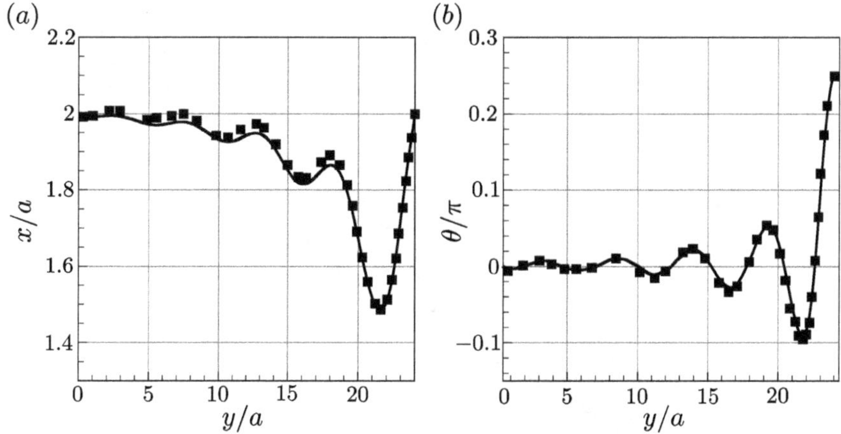

Figure 10.12 Location of center of mass (a) and orientation (b) for the elliptic particle sedimenting in a confined channel with $Re_t = 12.5$, $Fr_t = 0.126$. The present results (lines) are compared with finite-element numerical results of Xia et al. (2009) (symbols).

10.9 Fluttering and Tumbling of a Falling Plate

The same setup as in Figure 10.10 can be used to investigate a different problem: the motion of a freely falling plate, which transitions from fluttering to tumbling depending on the flow parameters. In fact, experimental studies by Belmonte et al. (1998) have shown that the problem is governed by two nondimensional quantities, namely the Reynolds Re and Froude Fr numbers which, for a body of aspect ratio $a/b = 8$ and mass $M = \pi a b \rho_s$, with a characteristic velocity $U_0 = \sqrt{2(\rho_s/\rho_f - 1)bg}$, read

$$Re = \frac{U_0 a}{\nu}, \qquad Fr = \sqrt{\frac{M}{\rho_f a^2}}, \tag{10.7}$$

where ρ_s and ρ_f are the mass densities of the body and of the fluid, respectively. In this example, a computational domain $30a \times 30a$ discretized by a uniform mesh of 2300×2300 nodes is adopted; on the top and side boundaries, free-slip conditions are imposed, while the bottom wall is no-slip. Along the plate boundary, Lagrangian markers have been evenly spaced at a distance $0.01a$, which is about 0.77 times the size of the homogeneous Eulerian mesh. A constant time step $\Delta t U_0/a = 0.005$ has been used for all cases.

In the case of fluttering dynamics, after an initial transient period, the plate oscillates while falling vertically without showing a net horizontal drift; accordingly, the initial position of the particle centroid is $(15a, 28a)$, again $(0, 0)$ being the bottom left corner of the domain. For the initial orientation of the

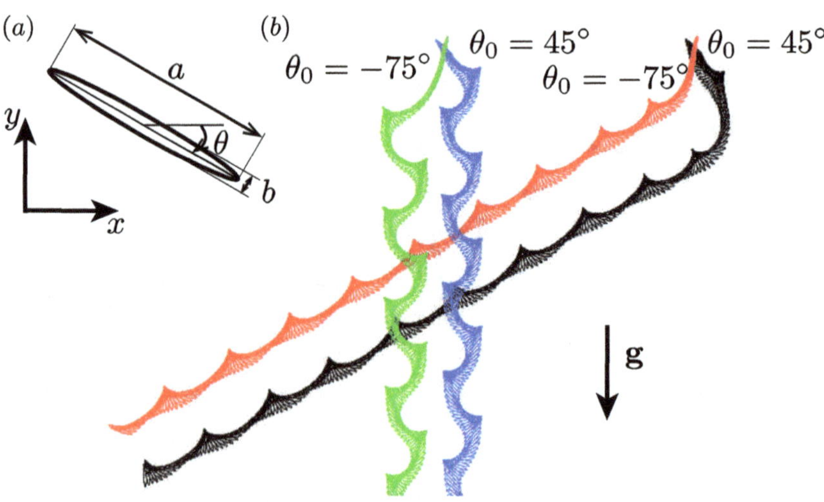

Figure 10.13 (a) Geometrical parameters for the plate falling in a quiescent fluid. (b) Overlapping of plate positions for fluttering (Re = 140, Fr = 0.45) and tumbling (Re = 420, Fr = 0.89) plates. Black pattern: tumbling case, $\theta_0 = 45°$; red pattern: tumbling case, $\theta_0 = -75°$; blue pattern: fluttering case, $\theta_0 = 45°$; green pattern: fluttering case, $\theta_0 = -75°$.

plate, two different values $\theta = 45°$ and $-75°$ have been simulated in order to verify that the long-term dynamics does not depend on the initial conditions. Following the study of Wan et al. (2012) fluttering conditions are obtained for $Re = 140$ and $Fr = 0.45$ and the same parameters have been adopted here.

On the other hand, in tumbling falling mode, the plate shows a mean horizontal drift; therefore, for a leftward migration, the initial position of the particle centroid is $(25a, 28a)$, which is in the top right corner region. Also in this case, the two initial orientations $\theta = 45°$ and $-75°$ have been simulated and the values $Re = 420$, $Fr = 0.89$ have been employed to replicate the results of Wan et al. (2012).

During the motion, pressure and viscous actions are exerted on the plate wet surface whose resultant is a time-dependent vector with components (F_x, F_y); their time evolution is presented in terms of force coefficients defined as $C_i = 2F_i/(\rho_f U_0^2 L)$ (note that in two-dimensional flows the loads are forces per unit width). The present results show very good agreement with those of Wan et al. (2012) obtained within the same conditions; here we show the sequence of overlapping instantaneous positions (Figure 10.13), force coefficients and velocity components for the various cases (Figures 10.14 and 10.15).

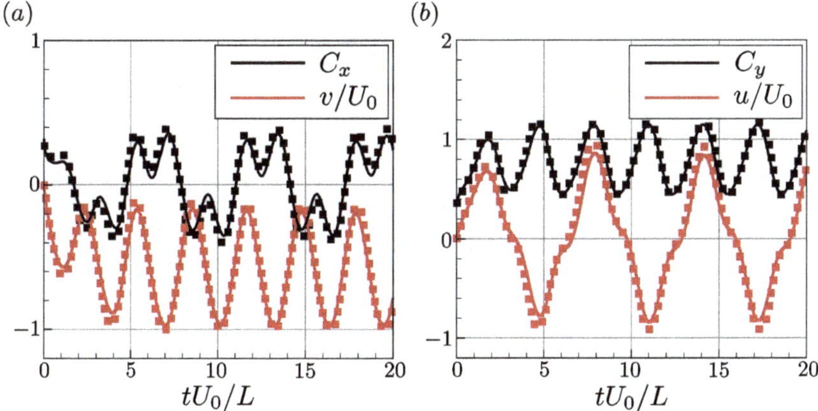

Figure 10.14 Fluttering case with $Re = 140$, $Fr = 0.45$, $\theta_0 = 45°$. (a) Horizontal force coefficient, C_x, and vertical velocity component, v. (b) Vertical force coefficient, C_y, and horizontal velocity component, u. Continuous lines indicate present results, while symbols indicate numerical results of Wan et al. (2012).

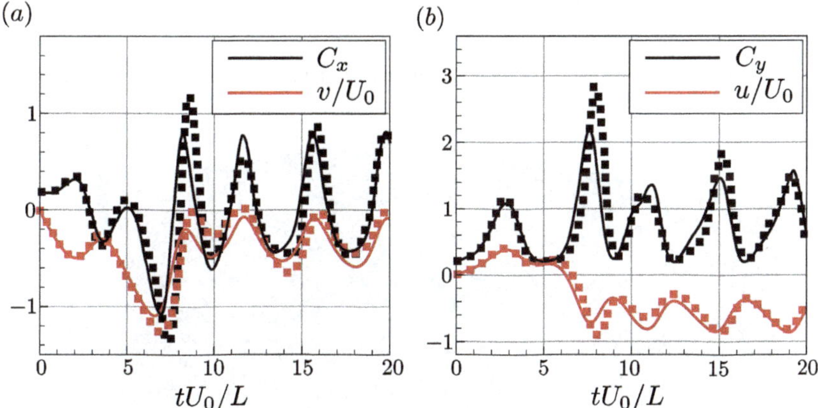

Figure 10.15 Tumbling case with $Re = 420$, $Fr = 0.89$, $\theta_0 = 45°$. (a) Horizontal force coefficient, C_x, and vertical velocity component, v. (b) Vertical force coefficient, C_y, and horizontal velocity component, u. Continuous lines indicate present results, while symbols indicate numerical results of Wan et al. (2012).

10.10 Flow Past a Sphere

We now compute the flow past a sphere of diameter D with uniform inflow streamwise velocity, U. The Reynolds number is set to $Re = UD/\nu = 100$ corresponding to a steady regime. We wish to point out that the flow around a sphere is a strict test case for IBMs since the boundary has all possible orientations with respect to the coordinate lines and any separations (if occurring) are

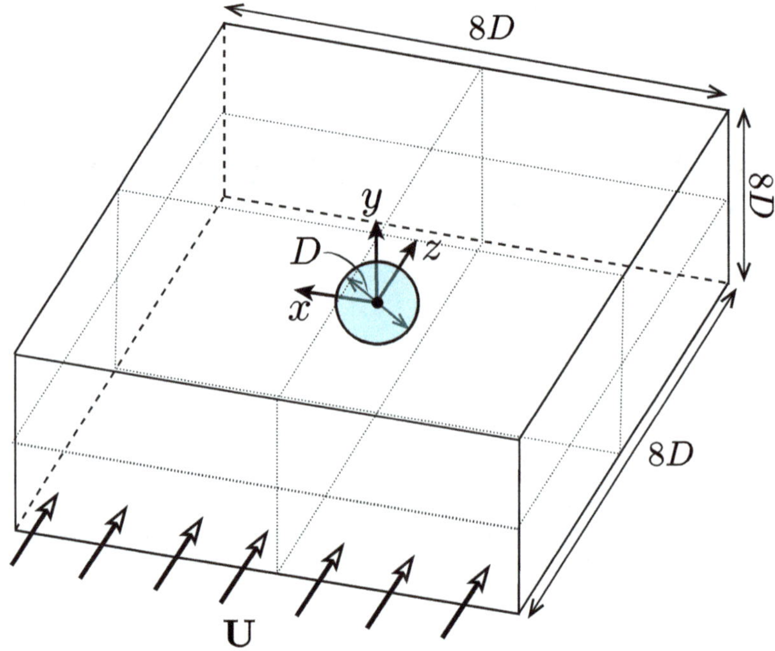

Figure 10.16 Schematic of the computational setup for the flow past a fixed sphere.

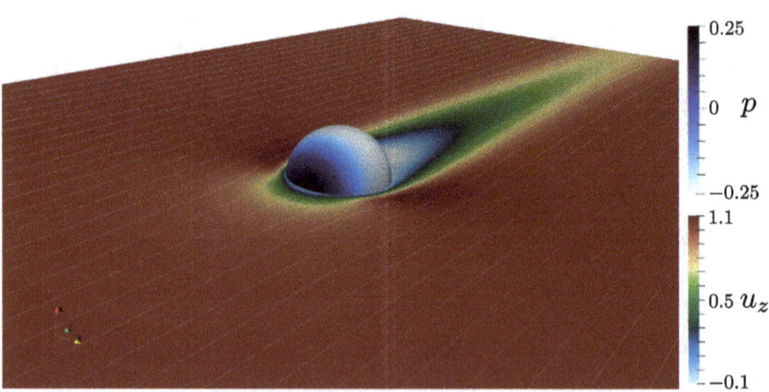

Figure 10.17 Flow around a fixed sphere at $Re = 100$. The planar section shows the streamwise velocity with superimposed velocity streamlines. The isocontours on the sphere show the surface pressure exerted by the flow.

determined by viscous phenomena rather than geometry discontinuities (edges or sharp corners).

As shown in Figure 10.16, the computational domain is $H_x \times H_y \times H_z = 8D \times 8D \times 8D$ with periodic conditions applied in all directions except for the flow direction z, where the uniform flow U is imposed at the inlet (south wall) and a

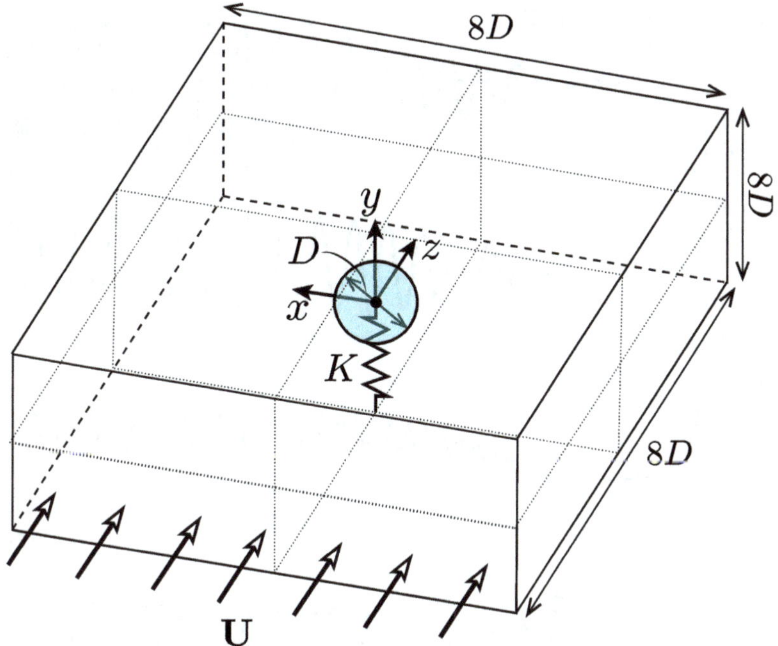

Figure 10.18 Schematic of the computational setup for the transverse flow-induced vibrations of an elastically mounted sphere in a uniform stream.

radiative condition is imposed at the outlet (north). The sphere is positioned at the center of the domain. A uniform grid of $N_x \times N_y \times N_z = 501 \times 501 \times 501$ nodes is used, corresponding to grid spacing of $0.016D$. The Lagrangian markers on the sphere surface are distributed uniformly, with a spacing of $0.01D$. A constant time step of $\Delta t = 5 \times 10^{-4} D/U$ is used.

The streamwise velocity within a symmetry plane of the sphere is shown in Figure 10.17 together with the in-plane flow streamlines. The blue isocontours on the sphere surface depict the surface pressure distribution. The corresponding drag coefficient is $C_D = 1.099$ using Eulerian IB (IFAD=1) and $C_D = 1.109$ with MLS Lagrangian IB (IMLS=1), which is in good agreement with the literature (for instance, within ≈ 2 percent the results of Fornberg (1988)). Given the steady character of the flow and the symmetry of the streamlines, all other force coefficients must be zero and this is confirmed by the code, which returns values within the round-off error.

10.11 Flow-Induced Vibrations of a Sphere

In this test case, we consider the flow-induced vibration of an elastically mounted sphere that is constrained to move in a direction transverse to the free stream (Rajamuni et al., 2018). The problem setup is sketched in Figure

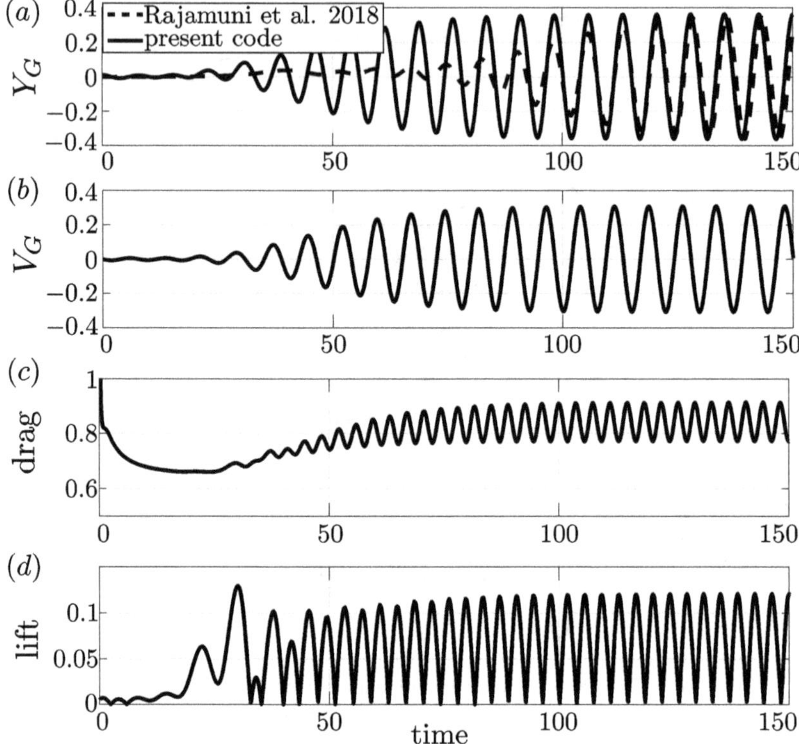

Figure 10.19 Nondimensional sphere (a) displacement Y_G, (b) velocity V_G, (c) hydrodynamic drag and (d) lift against nondimensional time. The dashed line in (a) corresponds to the results of Rajamuni et al. (2018).

10.18 with a sphere of diameter D that is mounted elastically on a spring of elastic constant K (no dashpot is applied) and its motion is restricted to the vertical y-direction, which is the only degree of freedom of the sphere.

The same computational setup of the previous test case is used and, similarly to the oscillating cylinders of Section 10.7, the position of the centroid of the sphere is given by equation (10.4). In order to anticipate the onset of the instability, the sphere is initially displaced by 1 percent D in the y-direction with respect to the neutral position of the spring $(0, 0, 0)$. The spring constant is set by the reduced velocity through equation (10.6), whereas the other mechanical parameter is the ratio between the density of the solid ρ_s and of the fluid ρ_f. Following Rajamuni et al. (2018), we focus on the case with $U_{\text{red}} = 7$ and mass ratio $\rho_s/\rho_f = 2.865$ (reduced mass $M_{\text{red}} = M/(\rho_f D^3) = \pi/6\rho_s/\rho_f = 1.5$), which manifests large flow-induced vibrations. The Reynolds number is $UD/\nu = 300$.

Figure 10.19 shows the nondimensional transversal displacement of the sphere, along with the corresponding velocity and the instantaneous drag

Figure 10.20 Snapshot of the vortical structures (Q-criterion) behind the oscillating sphere. The isosurface is color-mapped by velocity magnitude, with a legend indicating the corresponding values. The vertical plane shows the pressure field.

and lift (all quantities are made nondimensional with U and D). In this setting, the sphere oscillation frequency is synchronized with the natural frequency of the system, and it is close to the static body vortex shedding frequency. This locked-in regime with large-amplitude vortex-induced vibration is called *branch A*. The time history diagrams show that after several oscillation periods the system reaches an asymptotic state where it oscillates in a sinusoidal fashion symmetrically about its initial position. At this regime, the oscillating frequency and amplitude of the sphere agree well with the results of Rajamuni et al. (2018) (see Figure 10.19a). On the other hand, the difference in the transient phase can be ascribed to the destabilization mechanism of the sphere that can reorient the wake in any transversal direction, thus affecting the initial flow regime (see Figure 10.20). As a final note, we recall that a different initial condition for the sphere has been used (1 percent D displacement of the sphere in the y-direction) and that the computational domain in Rajamuni et al. (2018) is larger ($100D$ in each direction) with Dirichlet condition imposed in the lateral direction, rather than periodic ones as done here.

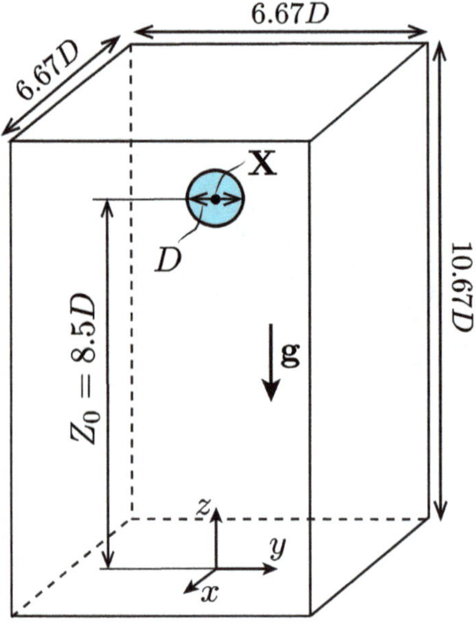

Figure 10.21 Schematic of the computational setup for the three-dimensional sphere settling under gravity in a closed container.

10.12 Settling of a Sphere in a Closed Tank

This benchmark aims at replicating some of the laboratory experiments performed by ten Cate et al. (2002) who investigated the motion of a rigid sphere falling under gravity in a closed container. In that study, particle image velocimetry was used to give accurate measurements of the sphere trajectory and velocity throughout the whole falling dynamics.

The flow parameters are such that the setup can be simulated under identical conditions, and a sketch is provided in Figure 10.21. Using the correlation of Abraham (1970) for the drag coefficient of a sphere in an unbounded domain $C_d = C_0(1 + \delta_0/\sqrt{Re})^2$ with $C_0\delta_0^2 = 24$ and $\delta_0 = 9.06$, a velocity scale can be computed as

$$U_\infty = \sqrt{\frac{4gD}{3C_d}(\gamma - 1)}, \tag{10.8}$$

where g is the gravitational acceleration, D is the sphere diameter and $\gamma = \rho_s/\rho_f$ is the solid-to-fluid density ratio.

Re	γ	U_∞ (m/s)	Fr	$\Delta t U_\infty / D$
1.5	1.155	0.038	0.0991	0.0001
4.1	1.161	0.060	0.156	0.0005
11.6	1.164	0.091	0.237	0.0010
31.9	1.167	0.128	0.334	0.0020

Table 10.1 *Run parameters for a rigid sphere falling in a closed container.*

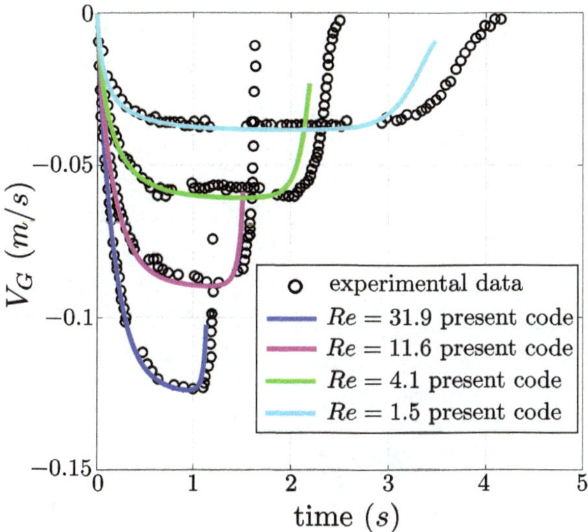

Figure 10.22 Sphere sedimentation velocity at Reynolds number 31.9 (blue line), 11.6 (magenta line), 4.1 (green line) and 1.5 (cyan line). The symbols correspond to the experimental result of ten Cate et al. (2002).

Based on these variables, the two nondimensional governing parameters are computed as

$$Re = \frac{U_\infty D}{\nu}, \quad \text{and} \quad Fr = \frac{U_\infty}{\sqrt{gD}}, \tag{10.9}$$

with ν the kinematic viscosity of the fluid. Two cases are considered whose parameters are summarized in Table 10.1.

The computational domain, aimed at replicating the closed tank of ten Cate et al. (2002), has all no-slip boundaries and dimensions $6.67D \times 6.67D \times 10.67D$, in the x-, y- and z-directions, which have been discretized by a uniform mesh of $241 \times 241 \times 385$ nodes. The mesh size is $\approx 0.0277D$ and the initial position of the sphere centroid, for all cases, is $\mathbf{X} = (0, 0, 8.5D)$. In Figure 10.22

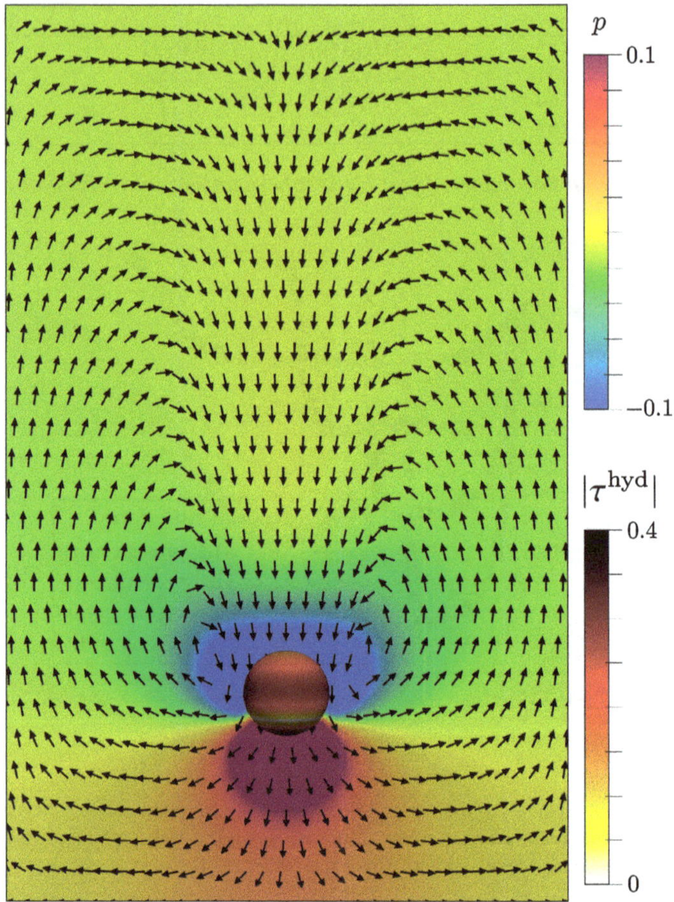

Figure 10.23 Snapshot of the pressure field and velocity directions during the falling phase of the sphere at $Re = 31.9$. The contours on the sphere indicate the distribution of hydrodynamic loads on its surface.

we report the sphere vertical velocity as a function of time, and the results are compared with the experimental data of ten Cate et al. (2002), obtaining a good agreement for all cases, without incorporating any lubrication or contact model. A snapshot of the instantaneous pressure field during the falling phase of the sphere is reported in Figure 10.23 along with the velocity field directions.

10.13 Three-Dimensional Flapping Flag

In this test case, we consider a full FSI three-dimensional flow around a "zero-thickness" deformable body subjected to a uniform flow. In a box of dimensions $[-2L, 2L] \times [-2.5L, 2.5L] \times [-2L, 3L]$, in the x-, y- and z-directions, respectively, a square flag of edge L is placed in the symmetry x–z-plane with the leading edge pinned at $y = z = 0$. The initial shape of the flag is that of a flat plate, inclined of $\theta = 0.1\pi$ with respect to the x–z-plane, see Figure 10.24. All the boundaries of the computational domain are periodic, except for the inflow $z = -2L$, where the velocity vector $\mathbf{U} = (0, 0, U)$, and for the outflow $z = 3L$, where velocity components are advected out of the domain with a nondimensional velocity `cou = 1` according to equation (10.2).

The simulation is performed on a uniform grid with $\Delta x = \Delta y = \Delta z = 0.02L$, which is achieved by a mesh of $201 \times 251 \times 251$ nodes. The flag is discretized by equilateral triangles of edges $\approx 0.0132L$. The solution is advanced in time using a constant nondimensional time step of $\Delta t U / L = 10^{-4}$.

The flow parameters are set to replicate the benchmark of Huang and Sung (2010) and other numerical results available from the literature. Specifically, gravity is not included among forces, the solid-to-fluid density ratio is $\gamma = \rho_s / \rho_f = 100$ and the Reynolds number $Re = UL/\nu = 200$. As in this case, a deformable body is considered, the elastic properties of the structure are part of the input; the flag thickness set in the structural solver is assumed $h = 0.01L$ so that its mass $M = \rho_s h L^2$ can be distributed among the nodes of the triangulated surface as described in Section 7.4.2. As a real flag is practically inextensible, in the simulation this is achieved by assigning a large Young modulus, which in nondimensional form becomes $\epsilon = Eh/(\rho_f U^2 L) = 1000$. The bending stiffness B of the structure is also prescribed in nondimensional form as $\beta = B/(\rho_f U^2 L^3) = 0.0001$.

It is worth mentioning that, in the method of interaction potentials, the elastic properties of the structure are set through the constants k_e and k_b which can be computed as described in Section 7.4.2. Figure 10.25 shows the time evolution of the y-coordinate of the trailing edge middle point, which indicates that the flag reaches a periodic flapping motion after an initial transient lasting about two periods. The results are in good agreement with the data of Huang and Sung (2010).

Study	Amplitude A/L	Strouhal number St
present	0.775	0.266
de Tullio and Pascazio (2016)	0.795	0.265
Tian et al. (2014)	0.812	0.263
Lee and Choi (2015)	0.752	0.265
Huang and Sung (2010)	0.780	0.262

Table 10.2 *Representative quantities for a flapping flag in uniform flow with*
$Re = 200$, $\gamma = 100$ and $\epsilon = 1000$ and $\beta = 0.0001$. Comparison of
peak-to-peak excursion amplitude, A/L, and the Strouhal number, St, for the
middle trailing-edge point.

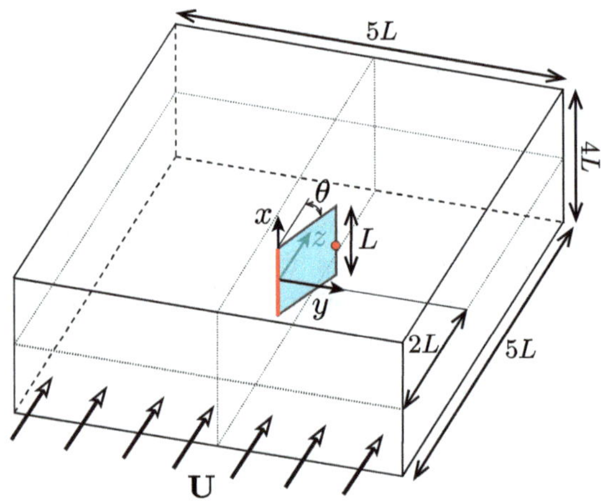

Figure 10.24 Schematic of the computational setup for the three-dimensional flapping flag in a uniform stream. The red thick segment indicates the pinned leading edge of the flag, while the red bullet on the trailing edge evidences the point where the time–dependent position has been sampled for analysis and comparison.

Table 10.2 reports the value of the oscillation amplitude A/L and Strouhal number $St = fL/U$, f being the frequency of the trailing-edge middle point, with results from the literature for an analogous flow.

Finally, the instantaneous vortical structures, identified by the Q-criterion around the flapping flag, are reported in Figure 10.25, showing the characteristic hairpin-like structure shed at each flapping, and in Figure 10.26.

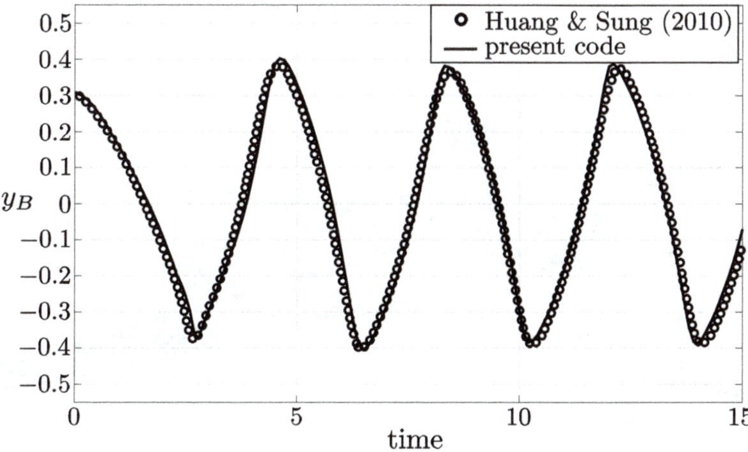

Figure 10.25 Time traces of the trailing-edge transverse location (middle point) of the flapping flag, for $Re = 200$, $\gamma = \rho_s/\rho_f = 100$ and $\beta = 0.0001$.

Figure 10.26 Vortical structures (Q-criterion) around the flapping flag.

10.14 Self-Propelled Manta Ray

As a final example, we consider the case of a swimming manta ray. Figure 10.27 shows the fish geometry together with the key geometric and kinematic quantities. BL and SL denote the body length and half the spanwise length, respectively, while A_u and A_d are the upward and downward half-amplitudes. The manta advances along the positive z-direction while flapping the pectoral fins in

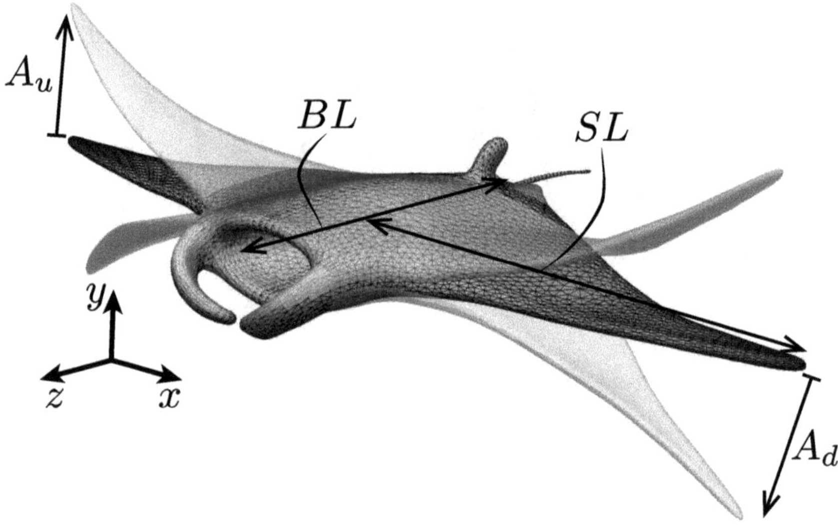

Figure 10.27 Manta ray geometry in the flat position and body frame coordinate system. The two shaded surfaces correspond to the maximum and minimum vertical deflection of the pectoral fins.

the vertical y-direction, with x being the wing span direction. The fin kinematics (Fish et al., 2016; Zhang et al., 2022a) in the body frame is imposed as

$$x(x_0, z_0, t) = \qquad x_0(1 - (1 - k)|f(z_0, t)|x_0/SL)\cos(f(z_0, t)\theta_{max}x_0/SL),$$
$$y(x_0, z_0, t) = y_0 + x_0(1 - (1 - k)|f(z_0, t)|x_0/SL)\sin(f(z_0, t)\theta_{max}x_0/SL),$$
$$z(x_0, z_0, t) = z_0,$$
$$\text{with } f(z_0, t) = \sin(\omega t - 2\pi W z_0/BL),$$

$$(10.10)$$

where $x(t), y(t), z(t)$ are the corresponding coordinates of the point in the initial position x_0, y_0, z_0 of the configuration with flat fins. Here we consider a symmetric amplitude motion of the manta $A_u = A_d = A = 0.35SL$ and the corresponding geometrical parameters $k = 0.9778$ and $\theta_{max} = 0.3661$ depending on A/SL have been determined as indicated in Zhang et al. (2022a). Moreover, we consider a flapping frequency $f = 1$ Hz (hence, $\omega = 2\pi$ Hz) with a chordwise traveling wavenumber $W = 6$.

The swimming kinematics (10.10) is imposed in the body frame, and the resulting flow patterns are obtained by solving the Navier–Stokes equations in the same frame of reference. The acceleration A_G and velocity V_G in the z-direction of the center of mass of the manta are thus given by

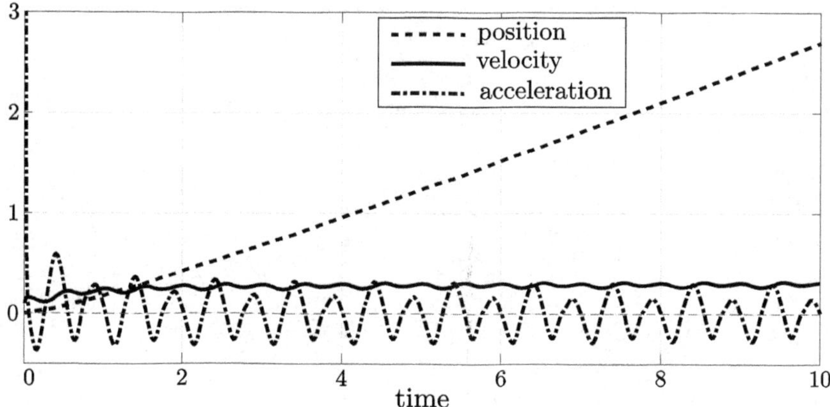

Figure 10.28 Self propulsion of the manta ray. Temporal variation of nondimensional acceleration A_G, velocity V_G and position Z_G of the center of mass.

$$MA_G = F_{p,z} + F_{v,z}$$
$$\frac{dV_G}{dt} = A_G, \quad \frac{dZ_G}{dt} = V_G, \tag{10.11}$$

where $F_{p,z}$ and $F_{v,z}$ are the resulting pressure and viscous forces and M is the mass of the manta-ray. The acceleration A_G corresponds to the translation inertial force entering in the body-frame Navier–Stokes equations (in the routine *calcnl3*), whereas the velocity V_G is used to update the Dirichlet condition at the south wall (routine *bouc_dq*).

Although the Reynolds number based on biological data (defined on the body length and swimming velocity) ranges between 5×10^4 and 5×10^5, we set here $Re = 5000$ in order to reduce the computational burden. Still, the main flow patterns and qualitative features are expected to be well captured even at this lower Reynolds number (Zhang et al., 2022a). Moreover, since the flow is solved in the reference frame translating with the manta, the computational domain can be made relatively narrow around the manta geometry and is equal to $H_z = 3SL$, $H_y = 1.5SL$ and $H_x = 3SL$, which is discretized with $241 \times 121 \times 241$ grid points. The corresponding dynamics is shown in Figure 10.28, where after an initial acceleration phase (about one time unit), the manta reaches a swimming regime with the acceleration oscillating between positive and negative values, whereas the velocity shows smaller oscillations around its mean. Hence, the corresponding displacement Z_G increases in a linear fashion. The flow patterns generated by the swimming manta are reported in Figure 10.29. The underlying grid (showing only a subset of the computational points) illustrates how the IBM enables the use of fixed, structured grids to solve flows around complex, moving objects.

 Numerical Examples

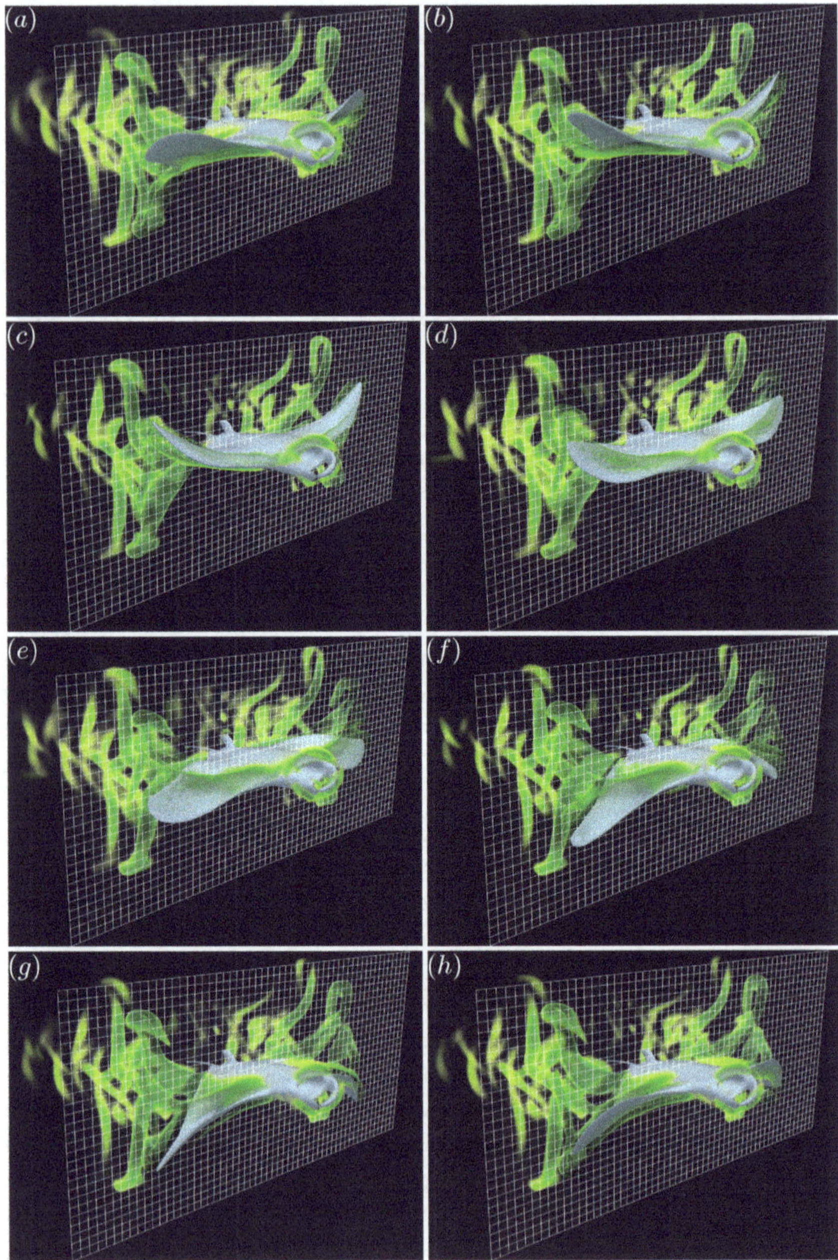

Figure 10.29 Instantaneous vorticity pattern (depicted by the Q-criterion, range 0–20) generated by a manta ray swimming through a fixed Eulerian grid. The eight snapshots are equispaced at $T/8$ within a period T.

References

Abraham, F. 1970. Functional dependence of drag coefficient of a sphere on Reynolds number. *Physics of Fluids*, **13**, 2194–2195.

Aftosmis, M. J., Berger, M., and Melton, J. E. 1998. Adaptive Cartesian mesh generation. In: Thompson, J., Soni, B., and Weatherill, N. (eds.), *Handbook of Grid Generation*. Vol. 57. CRC Press.

Aftosmis, M. J., Delanaye, M., and Haimes, R. 1999. Automatic generation of ready surface triangulations from CAD geometry. *AIAA Paper*, 1999–776.

Aftosmis, M. J., Berger, M. J., and Adomavicius, G. 2000. A parallel multilevel method for adaptively refines Cartesian grids with embedded boundaries. In: *AIAA Paper 2000-0808*.

Almgren, A. S., Nell, J. B., Colella, P., and Marthaler, T. 1997. A Cartesian grid projection method for the incompressible Euler equations in complex geometries. *SIAM, Journal of Scientific Computing*, **18**(5), 1289–1309.

Alva, G., Lin, Y., and Fang, G. 2018. An overview of thermal energy storage systems. *Energy*, **144**, 341.

Anderson, J. D. 1995. *Computational Methods for Fluid Dynamics*. McGraw-Hill.

Anderson, J. L. 1989. Colloidal transport by interfacial forces. *Annual Review of Fluid Mechanics*, **21**, 61–99.

Angot, P., Bruneau, C. H., and Frabrie, P. 1999. A penalization method to take into account obstacles in viscous flows. *Numerische Matematik*, **81**, 497–520.

Balaras, E. 2004. Modeling complex boundaries using an external force field on fixed Cartesian grids in large-eddy simulations. *Computers & Fluids*, **33**(3), 375–404.

Balaras, E., Benocci, C., and Piomelli, U. 1996. Two-layer approximate boundary conditions for large-eddy simulations. *AIAA Journal*, **34**, 1111–1119.

Basdevant, C., and Sadourny, R. 1983. Modélisation des échelles virtuelles dans la simulation numérique des écoulements turbulents tridimentionnels. *Journal Mécanique Théorique Appliqué*, 243–269.

Batchelor, G. K. 1982. *The Theory of Homogeneous Turbulence*. Cambridge University Press.

Beam, R. M. P., and Warming, R. F. 1976. An implicit finite-difference algorithm for hyperbolic systems in conservation-law form. *Journal of Computational Physics*, **22**, 87–110.

Beeson, M. 2012. Triangle tiling I: The tile is similar to ABC or has a right angle. *arXiv preprint arXiv:1206.2231*.

Belmonte, A., Elsenberg, H., and Moses, E. 1998. From flutter to tumble: inertial drag and Froude similarity in falling paper. *Physical Review Letters*, **81**, 345–348.

Benek, J. A., Steger, J. L., Dougherty, F. C., and Buning, P. G. 1986. *Chimera: A Grid-Embedding Technique*. Tech. rept. Report No. AEDC–85–64. Arnold Air Force Station.

Bernardini, M., Modesti, D., and Pirozzoli, S. 2016. On the suitability of the immersed boundary method for the simulation of high-Reynolds-number separated turbulent flows. *Computers & Fluids*, **130**, 84–93.

Beyer, R. P., and LeVeque, R. J. 1992. Analysis of a one-dimensional model for the immersed boundary method. *SIAM Journal on Numerical Analysis*, **29**(2), 332–364.

Bijl, H., Lucor, D., Mishra, S., and Schwab, C. 2013. *Uncertainty Quantification in Computational Fluid Dynamics*. Springer.

Blake, W. K. 1975. *Statistical Description of Pressure and Velocity Fields of the Trailing Edge of a Flat Surface*. Tech. rept. Report No. DTNSRDC–4241. David Taylor Naval Ship R. & D. Center, Bethesda, Maryland.

Blass, A., Zhu, X., Verzicco, R., Lohse, D., and Stevens, R. A. J. M. 2020. Flow organization and heat transfer in turbulent wall sheared thermal convection. *Journal of Fluid Mechanics*, **897**, A22.

Borazjani, I., and Sotiropoulos, F. 2009. Vortex induced vibrations of two cylinders in tandem arrangement in the proximity-wake interference region. *Journal of Fluid Mechanics*, **621**, 321–364.

Borazjani, I., Ge, L., and Sotiropoulos, F. 2008. Curvilinear immersed boundary method for simulating fluid structure interaction with complex 3D rigid bodies. *Journal of Computational Physics*, **227**, 7587–7620.

Brackbill, J. U., Kothe, D. B., and Zemach, C. 1992. A continuum method for modeling surface tension. *Journal of Computational Physics*, **100**, 335–354.

Briscolini, M., and Santangelo, P. 1989. Development of the mask method for incompressible unsteady flow. *Journal of Computational Physics*, **84**, 57–75.

Cabot, W., and Moin, P. 2000. Approximate wall boundary conditions in the large-eddy simulation of high Reynolds number flow. *iFlow, Turbulence and Combustion*, **63**, 269–291.

Cao, S., Chen, L., and Guo, R. 2023. Immersed virtual element methods for electromagnetic interface problems in three dimensions. *Mathematical Models and Methods in Applied Sciences*, **33**, 455–503.

Capizzano, F., Alterio, L., Russo, S., and de Nicola, C. 2019. A hybrid RANS-LES Cartesian method based on a skew-symmetric convective operator. *Journal of Computational Physics*, **390**, 359–379.

Carraturo, M., Kollmannsberger, S., Reali, A., Auricchio, F., and Rank, E. 2021. An immersed boundary approach for residual stress evaluation in selective laser melting processes. *Additive Manifacturing*, **46**, 102077.

Cenedese, C., and Straneo, F. 2023. Icebergs melting. *Annual Review of Fluid Mechanics*, **55**, 377–402.

Chen, W., Zou, S., Cai, Q., and Yang, Y. 2023. An explicit and non-iterative moving least-squares immersed-boundary method with low boundary velocity error. *Journal of Computational Physics*, **474**, 111803.

Chung, T. J. 2002. *Computational Fluid Dynamics*. Cambridge University Press.

Codony, D., Marco, O., Fernàndez-Mèndez, S., and Arias, I. 2019. An immersed boundary hierarchical B-spline method for flexoelectricity. *Computer Methods in Applied Mechanics and Engineering*, **354**, 750–782.

Coirier, W. J. 1994. *An Adaptively-Refined, Cartesian, Cell-Based Scheme for the Euler and Navier–Stokes Equations*. Tech. rept. NASA TM106754. NASA.

Cortez, R. 2000. A vortex/impulse method for immersed boundary motion in high Reynolds number flows. *Journal of Computational Physics*, **160**, 385–400.

Cortez, R., and Minion, M. 2000. The blob projection method for immersed boundary problems. *Journal of Computational Physics*, **161**(1), 428–453.

Cottet, H., and Maitre, E. 2004. A level-set formulation of immersed boundary methods for fluid–structure interaction problems. *Comptes Rendus Matematique*, **338**, 581–586.

Couston, L. A., Hester, E. W., Favier, B., Taylor, J. R., Holland, P. R., and Jenkins, A. 2021. Topography generation by melting and freezing in a turbulent shear flow. *Journal of Fluid Mechanics*, **911**, A44.

Cristallo, A., and Verzicco, R. 2006. Combined immersed boundary/large-eddy-simulations of incompressible three dimensional complex flows. *Flow, Turbulence and Combustion*, **77**(1–4), 3–26.

Curtis, S., Best, A., and Manocha, D. 2016. Menge: A modular framework for simulating crowd movement. *Collective Dynamics*, **1**(A1), 1–40.

de Graaf, J., Rempfer, G., and Holm, C. 2015. Diffusiophoretic self-propulsion for partially catalytic spherical colloids. *IEEE Transactions of Nanobiosciences*, **14**, 272–288.

de Tullio, M. D., and Pascazio, G. 2016. A moving-least-squares immersed boundary method for simulating the fluid-structure interaction of elastic bodies with arbitrary thickness. *Journal of Computational Physics*, **325**, 201–225.

de Tullio, M. D., De Palma, P., Iaccarino, G., Pascazio, G., and Napolitano, M. 2007. An immersed boundary method for compressible flows using local grid refinement. *Journal of Computational Physics*, **225**, 2098–2117.

Dihn, Q. V., Glowinski, R., He, J., Kwock, V., Pan, T. W., and Periaux, J. 1992. Lagrange multiplier approach to fictitious domain methods: Application to fluid dynamics and electromagnetics. Pages 151–194 of: *Proc. of V Int. Symp. Domain Decomp. Meth. for PDE*.

Durbin, P. 2021. *Advanced Approaches in Turbulence*. Elsevier.

Durbin, P., and Iaccarino, G. 2002. An approach to local refinement of structured grids. *Journal of Computational Physics*, **181**(2), 639–653.

Elghaoui, M., and Pasquetti, R. 1996. A spectral embedding method applied to the advection–diffusion equation. *Journal of Computational Physics*, **125**, 464–476.

Epstein, B., Luntz, A., and Nachson, A. 1992. Cartesian Euler methods for arbitrary aircraft configurations. *AIAA Journal*, **30**, 679–687.

Evans, M. W., and Harlow, F. H. 1957. *The Particle In Cell Method for Hydrodynamic Calculations*. Tech. rept. Report No. LAMS-2139. Los Alamos Scientific Laboratories.

Fadlun, E. A., Verzicco, R., Orlandi, P., and Mohd-Yusof, J. 2000. Combined immersed-boundary finite-difference methods for three-dimensional complex flow simulations. *Journal of Computational Physics*, **161**(1), 35–60.

Fedosov, D. A., Caswell, B., and Karniadakis, G. E. 2010. Systematic coarse-graining of spectrin-level red blood cell models. *Computer Methods in Applied Mechanics and Engineering*, **199**(29–32), 1937–1948.

Feng, Z. G., and Michaelides, E. 2004. The immersed boundary-lattice Boltzmann method for solving fluid–particles interaction problems. *Journal of Computational Physics*, **195**, 602–628.

Ferziger, J. H., and Peric, M. 2002. *Computational Methods for Fluid Dynamics*. Springer.

Fish, F. E., Schreiber, C. M., Moored, K. W., Liu, G., Dong, H., and Bart-Smith, H. 2016. Hydrodynamic performance of aquatic flapping: Efficiency of underwater flight in the manta. *Aerospace*, **3**(3), 20.

Fish, J., and Belytschko, T. 2007. *A First Course in Finite Elements*. Vol. 1. John Wiley & Sons Limited.

Fletcher, C. A. J. 1996. *Computational Techniques For Fluid Dynamics*. Vol. 1. Springer.

Fornberg, B. 1988. Steady viscous flow past a sphere at high Reynolds numbers. *Journal of Fluid Mechanics*, **190**, 471–489.

Forrer, H., and Jeltsch, R. 1998. A higher-order boundary treatment for Cartesian-grid methods. *Journal of Computational Physics*, **140**(2), 259–277.

Garland, R. 1982. *Microcomputers and Children in the Primary School*. Falmer Press.

Ge, L., and Sotiropoulos, F. 2007. A numerical method for solving the 3D unsteady incompressible Navier–Stokes equations in curvilinear domains with complex immersed boundaries. *Journal of Computational Physics*, **225**, 1782–1809.

Gentry, R. A., Martin, R. E., and Daly, B. J. 1966. An Eulerian differencing method for unsteady compressible flow problems. *Journal of Computational Physics*, **1**, 87–118.

Geraci, G., de Tullio, M. D., and Iaccarino, G. 2017. Polynomial chaos assessment of design tolerances for vortex-induced vibrations of two cylinders in tandem. *Artificial Intelligence for Engineering Design, Analysis and Manufacturing*, **31**, 185–198.

Germano, M. 1992. Turbulence: The filtering approach. *Journal of Fluid Mechanics*, **238**, 325–336.

Germano, M., Piomelli, U., Moin, P., and Cabot, W. H. 1991. A dynamic subgrid-scale eddy viscosity model. *Physics of Fluids*, **3**, 1760–1765.

Giannetti, F., and Luchini, P. 2007. Structural sensitivity of the first instability of the cylinder wake. *Journal of Fluid Mechanics*, **581**, 167–197.

Gilmanov, A., and Sotiropoulos, F. 2005. A hybrid Cartesian/immersed boundary method for simulating flows with 3D, geometrically complex, moving bodies. *Journal of Computational Physics*, **207**, 457–492.

Glowinski, R., Pan, T. W., and Périaux, J. 1998. Distributed Lagrange multiplier methods for incompressible viscous flow around moving rigid bodies. *Computational Methods Applied to Mechanical Engineering*, **151**, 181–194.

Goldstein, D., Handler, R., and Sirovich, L. 1993. Modeling a no-slip flow boundary with an external force field. *Journal of Computational Physics*, **105**, 354–366.

Goldstein, H., Pool, C., and Safko, J. 2001. *Classical Mechanics*, 3rd ed. Addison-Wesley.

Goldstine, H. H. 1980. *The Computer from Pascal to von Neumann*. Princeton University Press.

Golestanian, R., Liverpool, T., and Adjari, A. 2007. Designing phoretic micro- and nano-swimmers. *New Journal of Physics*, **9(5)**, 126.

Gonzalez, O., and Stuart, A. M. 2008. *A First Course in Continuum Mechanics*. Vol. 42. Cambridge University Press.

Griffiths, R. W. 2000. The dynamics of lava flows. *Annual Review of Fluid Mechanics*, **32**, 477–518.

Gsell, S., and Favier, J. 2021. Direct-forcing immersed-boundary method: A simple correction preventing boundary slip error. *Journal of Computational Physics*, **435**.

Guilmineau, E., and Queutey, P. 2002. A numerical simulation of vortex shedding from an oscillating circular cylinder. *Journal of Fluids and Structures*, **16(6)**, 773–794.

Guy, R. D., and Hartenstine, D. A. 2010. On the accuracy of direct forcing immersed boundary methods with projection methods. *Journal of Computational Physics*, **229**(7), 2479–2496.

Hammer, P. E., Sacks, M. S., Pedro, J., and Howe, R. D. 2011. Mass-spring model for simulation of heart valve tissue mechanical behavior. *Annals of Biomedical Engineering*, **39**(6), 1668–1679.

Harlow, F. H., and Whelch, J. E. 1971. Numerical calculation of time-dependent viscous incompressible flow of fluid with free surface. *Physics of Fluids*, **8(12)**, 2182–2189.

Hartmann, E. 1998. A marching method for the triangulation of surfaces. *Visual Computer*, **14**, 95–108.

Heinrich, J. C., and Vionnet, C. A. 1995. The penalty method for the Navier–Stokes equations. *Archives of Computational Methods in Engineering*, **2**, 51–65.

Hester, E. W., Couston, L. A., Favier, B., Burns, K. J., and Vasil, G. M. 2020. Improved phase–field models of melting and dissolution in multi-component flows. *Proceedings of Royal Society A*, **476**, 20200508.

Hirt, C. W., and Nichols, B. D. 1981. Volume of fluid (VOF) method for the dynamics of free boundaries. *Journal of Computational Physics*, **39**, 201–225.

Hoefnagels, P. B. J., Wei, P., Narezo Guzman, D., Sun, C., Lohse, D., and Ahlers, G. 2017. Large-scale flow and Reynolds numbers in the presence of boiling in locally heated turbulent convection. *Physical Review Fluids*, **2**, 074604.

Holzapfel, G. A. 2002. *Nonlinear Solid Mechanics: A Continuum Approach for Engineering Science*. Kluwer Academic Publishers.

Huang, W. X., and Sung, H. J. 2010. Three-dimensional simulation of a flapping flag in a uniform flow. *Journal of Fluid Mechanics*, **653**, 301–336.

Iaccarino, G., and Verzicco, R. 2003. Immersed boundary technique for turbulent flow simulations. *Applied Mechanics Reviews*, **56**, 331–347.

Irons, B., and Tuck, R. C. 1969. A version of the Aitken acceleration for computer implementation. *International Journal for Numerical Methods in Engineering*, **1**, 275–277.

Kalitzin, G. 1997. Validation and development of two-equation turbulence models. Validation of CFD codes and assessment of turbulence models. In: Haase, W. et al. (eds.), *Notes on Numerical Fluid Mechanics Series*. Vol. 57. Vieweg.

Kalitzin, G., and Iaccarino, G. 2002. *Turbulence Modeling in an Immersed-Boundary RANS Method*. Tech. rept. CTR Annual Research Briefs, NASA Ames/Stanford University, 1997. CTR/NASA.

Kalitzin, G., Iaccarino, G., and Khalighi, B. 2003. Towards an immersed boundary solver for RANS simulations. *AIAA Paper*, 2003–770.

Kalitzin, G., Medic, G., Iaccarino, G., and Durbin, P. 2005. Near-wall behavior of RANS turbulence models and implications for wall functions. *Journal of Computational Physics*, **204**, 265–291.

Kang, S. 2015. An improved near-wall modeling for large-eddy simulation using immersed boundary methods. *International Journal Numerical Methods in Fluids*, **78**, 76–88.

Kang, S., Iaccarino, G., and Moin, P. 2009. Accurate immersed-boundary reconstructions for viscous flow simulations. *AIAA Journal*, **47**(7), 1750–1760.

Kantor, Y., and Nelson, D. R. 1987. Phase transitions in flexible polymeric surfaces. *Physical Review A*, **36**(8), 4020.

Kempe, T., and Fröhlich, J. 2012. An improved immersed boundary method with direct forcing for the simulation of particle laden flows. *Journal of Computational Physics*, **231**(9), 3663–3684.

Khandra, K., Angot, P., Parneix, S., and Caltagirone, J. P. 2000. Fictitious domain approach for numerical modelling of Navier–Stokes equations. *International Journal of Numerical Methods in Fluids*, **34**, 651–684.

Kim, J., and Moin, P. 1985. Application of a fractional-step method to incompressible Navier–Stokes equations. *Journal of Computational Physics*, **59**, 308–323.

Kim, J., Kim, D., and Choi, H. 2001. An immersed-boundary finite-volume method for simulations of flow in complex geometries. *Journal of Computational Physics*, **171**, 60–85.

Kirkpatrick, M. P., Armfield, S. W., and Kent, J. H. 2003. A representation of curved boundaries for the solution of the Navier–Stokes equations on a staggered three-dimensional Cartesian grid. *Journal of Computational Physics*, **184**(1), 1–36.

Kuznetsov, Y. A. 2001. Domain decomposition and fictitious domain methods with distributed Lagrange multipliers. Pages 67–75 of: *Proc. of XIII Int. Conf. Domain Decomp. Meth.*

Kwak, D., Chang, J. L. C., Shanks, S. P., and Chakravarthy, S. R. 1986. A three-dimensional incompressible Navier–Stokes flow solver using primitive variables. *AIAA Journal*, **24**(3), 390–396.

Larsson, J., Kawai, S., Bodart, J., and Bermejo-Moreno, I. 2016. Large eddy simulation with modeled wall-stress: Recent progress and future directions. *Mechanical Engineering Reviews*, **3**, 00418.

Launder, B. E., and Sandham, N. D. 2002. *Closure Strategies for Turbulent and Transitional Flows*. Cambridge University Press.

Lax, P. D., and Richtmyer, R. D. 1956. Survey of the stability of linear finite difference equations. *Communications in Pure and Applied Mathematics*, **9**, 267–293.

Lee, H., and Choi, H. 2015. A discrete-forcing immersed boundary method for the fluid-structure interaction of an elastic slender body. *Journal of Computational Physics*, **309**, 529–546.

Lee, H., and Liu, Y. 2023. A ghost-point based second order accurate finite difference method on uniform orthogonal grids for electromagnetic scattering around curved perfect electric conductors with corners. *Journal of Computational Physics*, **490**, 112314.

Li, J., Dao, M., Lim, C. T., and Suresh, S. 2005. Spectrin-level modeling of the cytoskeleton and optical tweezers stretching of the erythrocyte. *Biophysical Journal*, **88**(5), 3707–3719.

Lilly, D. K. 1992. A proposed modification of the Germano subgrid-scale closure method. *Physics of Fluids*, **4**, 633–635.

Liu, G.-R., and Gu, Y.-T. 2005. *An Introduction to Meshfree Methods and their Programming*. Springer Science & Business Media.

Liu, H. R., Ng, C. S., Chong, K. L., Lohse, D., and Verzicco, R. 2021. An efficient phase–field method for turbulent multiphase flows. *Journal of Computational Physics*, **446**, 110659.

Lorstadt, D., Francois, M., Shyy, W., and Fuchs, L. 2004. Assessment of volume of fluid and immersed boundary methods for droplet computations. *International Journal for Numerical Methods in Fluids*, **46**, 109–125.

Louda, P., Kozel, K., and Prihoda, J. 2008. Numerical solution of 2D and 3D viscous incompressible steady and unsteady flows using artificial compressibility method. *International Journal for Numerical Methods in Fluids*, **56**, 1399–1407.

Lu, R., Yu, P., and Sui, Y. 2024. A computational study of cell membrane damage and intracellular delivery in a cross-slot microchannel. *Soft Matter*, **20**, 4057–4071.

Luo, A. A., Sachdev, A. K., and Apelian, D. 2022. Alloy development and process innovations for light metals casting. *Journal of Materials Processing Technology*, **306**, 117606.

Lupo, G., Gruber, A., Brandt, L., and Duwig, C. 2020. Direct numerical simulation of spray droplet evaporation in hot turbulent channel flow. *International Journal of Heat and Mass Transfer*, **160**, 120184.

Mahesh, K., and Moin, P. 1998. Direct numerical simulation: A tool in turbulence research. *Annual Review of Fluid Mechanics*, **30**, 539–578.

Mathieu, J., and Scott, J. 2000. *An Introduction to Turbulent Flow*. Cambridge University Press.

McQueen, D. M., and Peskin, C. S. 1989. A three-dimensional computational method for blood flow in the heart II. Contractile fibers. *Journal of Computational Physics*, **82**, 289–297.

Menter, F. M. 1993. Zonal two-equation k–ω turbulence models for aerodynamic flows. *IAA Paper*, 2906–93.

Mittal, R., and Iaccarino, G. 2005. Immersed boundary methods. *Annual Review of Fluid Mechanics*, **37**, 239–261.

Mittal, R., and Moin, P. 1997. Suitability of upwind-biased finite difference schemes for large-eddy simulation of turbulent flows. *AIAA Journal*, **35**, 1435–1437.

Mohd-Yusof, J. 1997. *Combined Immersed Boundaries/B-Splines Methods for Simulations of Flows in Complex Geometries*. Tech. rept. CTR Annual Research Briefs, NASA Ames/Stanford University, 1997. CTR/NASA.

Mok, D. P., Wall, W. A., and Ramm, E. 2001. Accelerated iterative substructuring schemes for instationary fluid–structure interaction. *Computational Fluid and Solid Mechanics*, 1325–1328.

Nealen, A., Müller, M., Keiser, R., Boxerman, E., and Carlson, M. 2006. Physically based deformable models in computer graphics. Pages 809–836 of: *Computer Graphics Forum*, vol. 25(4). Wiley Online Library.

Nyquist, H. 1928. Certain topics in telegraph transmission theory. *Transactions of the AIEE.*, **47**(2), 617–644.

Orlandi, P. 2012. *Fluid Flow Phemomena: A Numerical Toolkit*. Vol. 55. Springer Science & Business Media.

Orlandi, P., and Leonardi, S. 2006. DNS of turbulent channel flows with two- and three-dimensional roughness. *Journal of Turbulence*, N73.

O'Rourke, J. 1998. *Computational Geometry in C*. Cambridge University Press.

Pember, R. B., Bell, P., Colella, P., Crutchfield, W. Y., and Welcome, M. L. 1995. An adaptive Cartesian grid method for unsteady compressible flow in irregular regions. *Journal of Computational Physics*, **120**, 278–304.

Peskin, C. S. 1972a. Flow patterns around heart valves: A numerical method. *Journal of Computational Physics*, **10**, 252–271.

Peskin, C. S. 1972b. Numerical analysis of blood flow in the heart. *Journal of Computational Physics*, **25**, 220–252.

Peskin, C. S. 2002. The immersed boundary method. *Acta Numerica*, 479–517.

Peskin, C. S., and McQueen, D. M. 1989. A three-dimensional computational method for blood flow in the heart I. Immersed elastic fibers in a viscous incompressible fluid. *Journal of Computational Physics*, **81**, 372–405.

Piomelli, U. 2008. Wall-layer models for large-eddy simulations. *Progress in Aerospace Science*, **44**, 437–446.

Pitsch, H., and Duchamp de Lageneste, L. 2002. Large-eddy simulation of premixed turbulent combustion using a level-set approach. *Proceedings of the Combustion Institute*, **29**(2), 2002–2008.

Pope, S. B. 2000. *Turbulent Flows*. Cambridge University Press.

Popinet, S. 2003. Gerris: A tree-based adaptive solver for the incompressible Euler equations in complex geometries. *Journal of Computational Physics*, **190**(2), 572–600.

Prosperetti, A., and Oğuz, H. N. 2001. Physalis: A new $o(N)$ method for the numerical simulation of disperse systems: Potential flow of spheres. *Journal of Computational Physics*, **167**, 196–216.

Prouvost, L., Belme, A., and Fuster, D. 2024. A metric-based adaptive mesh refinement criterion under constrain for solving elliptic problems on quad/octree grids. *Journal of Computational Physics*, **506**, 112941.

Quadrio, M., Frohnapfel, B., and Hasegawa, Y. 2016. Does the choice of the forcing term affect flow statistics in DNS of turbulent channel flow? *European Journal of Mechanics B/Fluids*, **55**(2), 286–293.

Rai, M., and Moin, P. 1991. Direct simulations of turbulent flow using finite-difference schemes. *Journal of Computational Physics,* **96**(1), 15–53.

Rajamuni, M. M., Thompson, M. C., and Hourigan, K. 2018. Transverse flow-induced vibrations of a sphere. *Journal of Fluid Mechanics*, **837**, 931–966.

Rich, M. 1963. *A Method for Eulerian Fluid Dynamics,*. Tech. rept. Report No. LAMS-2826. Los Alamos Scientific Laboratories.

Roma, A. M., Peskin, C. S., and Berger, M. J. 1999. An adaptive version of the immersed boundary method. *Journal of Computational Physics*, **153**(2), 509–534.

Saiki, E. M., and Biringen, S. 1996. Numerical simulations of a cylinder in uniform flow: Application of a virtual boundary method. *Journal of Computational Physics*, **123**, 450–465.

Sapahi, F., Verzicco, R., Lohse, D., and Krug, D. 2024. Mass transport at gas-evolving electrodes. *Journal of Fluid Mechanics*, **983**, A19.

Saulev, V.K. 1963. On the solution of some boundary value problems on high performance computers by fictitious domain method. *Siberian Mathematical Journal*, **4**, 912–925.

Schillinger, D., Harari, I., Hsu, M.-C., Kamensky, D., Stoter, S. K. F., Yu, Y., and Zhao, Y. 2016. The non-symmetric Nitsche method for the parameter-free imposition of weak boundary and coupling conditions in immersed finite elements. *Computer Methods in Applied Mechanics and Engineering*, **309**, 625–652.

Schlichting, H. 1968. *Boundary Layer Theory*. McGraw-Hill.

Schmidt, M. 1993. Cutting cubes – visualizing implicit surfaces by adaptive polygonization. *Visual Computer*, **10**, 101–115.

Shannon, C. E. 1948. A mathematical theory of communication. *Bell System Technical Journal*, **27**(3), 379–423.

Smagorinsky, J. 1963. General circulation experiments with the primitive equations: I. The basic experiment. *Monthly Weather Review*, **91**, 99–164.

Smolinski, M. S., Hamburg, M. A., and Lederberg, J. 2003. *Microbial Threats to Health: Emergence, Detection, and Response*. The National Academies Press.

Sotiropoulos, F., and Yang, X. 2014. Immersed boundary methods for simulating fluid–structure interaction. *Progress in Aerospace Sciences*, **65**, 1–21.

Spandan, V., Lohse, D., de Tullio, M. D., and Verzicco, R. 2018. A fast moving least squares approximation with adaptive Lagrangian mesh refinement for large scale immersed boundary simulations. *Journal of Computational Physics*, **375**, 228–239.

Su, S.-W., Lai, M.-C., and Lin, C.-A. 2007. An immersed boundary technique for simulating complex flows with rigid boundary. *Computers & Fluids*, **36**(2).

Sun, L., Yao, W., Robinson, T., Simao, M., and Armstrong, C. 2020. A framework of gradient-based shape optimization using feature-based CAD parameterization. *AIAA Paper*, 2020–0889.

Sussman, M., and Puckett, E. G. 2000. A coupled level set and volume-of-fluid method for computing 3D and axisymmetric incompressible two-phase flows. *Journal of Computational Physics*, **162**, 301–337.

Tadmor, E. B., Miller, R. E., and Elliott, R. S. 2012. *Continuum Mechanics and Thermodynamics*. Cambridge University Press.

Taira, K., and Colonius, T. 2007. The immersed boundary method: A projection approach. *Journal of Computational Physics*, **225**, 2118–2137.

ten Cate, A., Nieuwstad, C. H., Derksen, J. J., and Van den Akker, H. E. A. 2002. Particle imaging velocimetry experiments and lattice-Boltzmann simulations on a single sphere settling under gravity. *Physics of Fluids*, **14**, 4012–4025.

Tennekes, H., and Lumley, J. L. 1972. *A First Course in Turbulence*. MIT Press.

Tessicini, F., Iaccarino, G., Fatica, M., Wang, M., and Verzicco, R. 2002. Wall modeling for large-eddy simulation using an immersed boundary method. *Center for Turbulence Research Annual Research Briefs*, **2002**, 181–187.

Thompson, J. F., Soni, B. K., and Wheatherhill, N. P. 1999. *Handbook of Grid Generation*. CRC Press.

Tian, F. B., Dai, H., Luo, H., Doyle, F. J., and Rousseau, B. 2014. Fluid–structure interaction involving large deformations: 3D simulations and applications to biological systems. *Journal of Computational Physics*, **258**, 451–469.

Tryggvason, G., Scardovelli, R., and Zaleski, S. 2011. *Direct Numerical Simulations of Gas–Liquid Multiphase Flows*. Cambridge University Press.

Tseng, Y. H., and Ferziger, J. H. 2003. A ghost-cell immersed boundary method for flow in complex geometry. *Journal of Computational Physics*, **192**, 593–623.

Udaykumar, H. S., Mittal, R., and Shyy, W. 1999. Computation of solid–liquid phase fronts in the sharp interface limit on fixed grids. *Journal of Computational Physics*, **153**(2), 535–574.

Udaykumar, H. S., Mittal, R., Rampunggoon, P., and Khanna, A. 2001. A sharp interface Cartesian grid method for simulating flows with complex moving boundaries. *Journal of Computational Physics*, **174**(1), 345–380.

Uhlmann, M. 2005. An immersed boundary method with direct forcing for the simulation of particulate flows. *Journal of Computational Physics*, **209**, 448–476.

Usyk, T. P., Mazhari, R., and McCulloch, A. D. 2000. Effect of laminar orthotropic myofiber architecture on regional stress and strain in the canine left ventricle. *Journal of Elasticity and the Physical Science of Solids*, **61**, 143–164.

Vagnoli, Giovanni, Scarpolini, Martino A, Verzicco, Roberto, and Viola, Francesco. 2025. A local and explicit forcing correction for Lagrangian immersed boundary methods. *Computer Physics Communications*, 109741.

van Driest, E. R. 1956. On turbulent flow near a wall. *Journal of the Aeronautical Science*, **23**, 1007–1011.

Van Gelder, A. 1998. Approximate simulation of elastic membranes by triangulated spring meshes. *Journal of Graphics Tools*, **3**(2), 21–41.

Vanella, M., and Balaras, E. 2009. A moving-least-squares reconstruction for embedded-boundary formulations. *Journal of Computational Physics*, **228**(18), 6617–6628.

Verzicco, R. 2002. Sidewall finite-conductivity effects in confined turbulent thermal convection. *Journal of Fluid Mechanics*, **473**, 201–210.

Verzicco, R. 2023. Immersed boundary methods: A historical perspective and a future outlook. *Annual Review of Fluid Mechanics*, **55**, 129–155.

Verzicco, R, and Orlandi, P. 1996. A finite-difference scheme for three-dimensional incompressible flows in cylindrical coordinates. *Journal of Computational Physics*, **123**(2), 402–414.

Verzicco, R., Orlandi, P., Mohd-Yusof, J., and Haworth, D. 2000. Large-eddy simulation in complex geometric configurations using boundary body forces. *AIAA Journal*, **38**, 427–433.

Verzicco, R., Fatica, M., Iaccarino, G., and Orlandi, P. 2004. Flow in an impeller-stirred tank using an immersed-boundary method. *AIChE Journal*, **50**, 1109–1118.

Viecelli, J. 1969. A method for including arbitrary external boundaries in the MAC incompressible fluid computing technique. *Journal of Computational Physics*, **4**, 543–551.

Viecelli, J. 1971. A computing method for incompressible flows bounded by moving walls. *Journal of Computational Physics*, **8**, 119–143.

Viola, F., Meschini, V., and Verzicco, R. 2020. Fluid–structure-electrophysiology interaction (FSEI) in the left-heart: A multi-way coupled computational model. *European Journal of Mechanics-B/Fluids*, **79**, 212–232.

Viola, F., Del Corso, G., and Verzicco, R. 2023a. High-fidelity model of the human heart: An immersed boundary implementation. *Physical Review Fluids*, **8**(10), 100502.

Viola, F., Del Corso, G., De Paulis, R., and Verzicco, R. 2023b. GPU accelerated digital twins of the human heart open new routes for cardiovascular research. *Scientific Reports*, **13**(1), 8230.

Wan, H., Dong, H., and Liang, Z. 2012. Vortex formation of freely falling plates. Pages 1–10 of: *Aerospace Exposition, Tennessee, Jan. 9–12, 2012*. American Institute of Aeronautics and Astronautics.

Wang, M., and Moin, P. 2000. Computation of trailing-edge flow and noise using large-eddy simulation. *AIAA Journal*, **38**, 2001–2009.

Wang, S., Vanella, M., and Balaras, E. 2019. A hydrodynamic stress model for simulating turbulence/particle interactions with immersed boundary methods. *Journal of Computational Physics*, **382**, 240–263.

Welch, J. E., Harlow, F. H., Shannon, J. P., and Daly, B. J. 1965. *The MAC Method: A Computing Technique for Solving Viscous, Incompressible, Transient Fluid-Flow Problems Involving Free Surfaces*. Tech. rept. Report No. LA-3425. Los Alamos Scientific Laboratories.

Wie, B. 1998. *Space Vehicle Dynamics and Control*. AIAA.

Wilcox, D. C. 1998. *Turbulence Modelling for CFD*, 2nd edn. DCW Industries, Inc.

Worster, M. G., Batchelor, G. K., and Moffat, H. K. 2000. Solidification of fluids. *Perspectives in Fluid Mechanics*, **742**, 393–446.

Xia, Z., Connington, K. W., Rapaka, S., Yue, P., Feng, J. J., and Chen, S. 2009. Flow patterns in the dedimentation of an elliptical particle. *Journal of Fluid Mechanics*, **625**, 249–272.

Yanenko, N. N. 1967. *Fractional-Step Methods for Multi-dimensional Solutions*. Nauka Press.

Yang, R., Howland, C., Liu, H. R., Verzicco, R., and Lohse, D. 2023a. Bistability in radiatively heated melt ponds. *Physical Review Letters*, **131**, 234002.

Yang, R., Howland, C., Liu, H. R., Verzicco, R., and Lohse, D. 2023b. Ice melting in salty water: Layering and non-monotonic dependence on the mean salinity. *Journal of Fluid Mechanics*, **969**, R2.

Yang, X. I. A., Sadique, J., Mittal, R., and Meneveau, C. 2015. Integral wall model for large eddy simulations of wall-bounded turbulent flows. *Physics of Fluids*, **27**, 025112.

Yang, Y., and Balaras, E. 2006. An embedded-boundary formulation for large-eddy simulation of turbulent flows interacting with moving boundaries. *Journal of Computational Physics*, **215**, 12–40.

Ye, T., Mittal, R., Udaykumar, H. S., and Shyy, W. 1999. An accurate Cartesian grid method for viscous incompressible flows with complex immersed boundaries. *Journal of Computational Physics*, **156**(2), 209–240.

Yildiran, I. N., Beratlis, N., Capuano, F., Loke, Y.-H., Squires, K., and Balaras, E. 2024. Pressure boundary conditions for immersed-boundary models. *Journal of Computational Physics*, **510**.

Zhang, D., Huang, Q.-G., Pan, G., Yang, L.-M., and Huang, W.-X. 2022a. Vortex dynamics and hydrodynamic performance enhancement mechanism in batoid fish oscillatory swimming. *Journal of Fluid Mechanics*, **930**, A28.

Zhang, W., Shahmardi, A., Choi, K., Tammisola, O., Brandt, L., and Mao, X. 2022b. A phase-field method for three-phase flows with icing. *Journal of Computational Physics*, **458**, 111104.

Zhang, Z., and Prosperetti, A. 2005. A second-order method for three-dimensional particle simulation. *Journal of Computational Physics*, **210**, 292–324.

Zhong, K., Howland, C. J., Lohse, D., and Verzicco, R. 2025. A front-tracking immersed-boundary framework for simulating Lagrangian melting problems. *Journal of Computational Physics,* **525**, 113762.

Zhu, X., Chen, Y., Chong, K. L., Lohse, D., and Verzicco, R. 2024. A boundary condition-enhanced direct-forcing immersed boundary method for simulations of three-dimensional phoretic particles in incompressible flows. *Journal of Computational Physics*, **509**, 113028.

Index